Ethnobotany in the Neotropics

Proceedings: Ethnobotany in the Neotropics Symposium
Society for Economic Botany
13–14 June, 1983
Oxford, Ohio, U.S.A.

**Edited by G. T. Prance
& J. A. Kallunki**

**The New York Botanical Garden
Bronx, New York, U.S.A.
1984**

Volume 1, ADVANCES IN ECONOMIC BOTANY

This is Volume 1 of ADVANCES IN ECONOMIC BOTANY

Copyright © 1984 The New York Botanical Garden

Published by The New York Botanical Garden, Bronx, New York 10458, U.S.A.

Issued 18 September 1984

Library of Congress Cataloging in Publication Data

Ethnobotany in Neotropics Symposium (1983 : Oxford, Ohio)
Ethnobotany in the neotropics.

(Advances in economic botany ; v. 1)
Includes bibliographies.
1. Indians of South America—Amazon River Region—
Ethnobotany—Congresses. 2. Ethnobotany—Amazon River
Region—Congresses. 3. Ethnobotany—Latin America—
Congresses. 4. Ethnobotany—Tropics—Congresses.
I. Prance, Ghillean T., 1937–   . II. Kallunki, J. A.
(Jacquelyn A.) III. New York Botanical Garden.
IV. Title.   V. Series.
F2519.1.A6E87 1984   581.6'1'09811   84-16517

ISBN 0-89327-253-1

Printed by Allen Press, Lawrence, Kansas, U.S.A.

Contributors

F. Ayala Flores, Universidad Nacional de la Amazonia Peruana, Iquitos, Peru

M. J. Balick, The New York Botanical Garden, Bronx, NY 10458

B. Berlin, Department of Anthropology and Center for Latin American Studies, University of California, Berkeley, CA 94720

J. S. Boster, Department of Anthroplogy, 211 Lafferty Hall, University of Kentucky, Lexington, KY 40506-0024

C. B. Heiser, Jr., Department of Biology, Indiana University, Bloomington, IN 47405

W. H. Lewis and Memory P. F. Elvin-Lewis, Department of Biology and School of Dental Medicine, Washington University, St. Louis, MO 63130

T. Plowman, Herbarium, Field Museum of Natural History, Roosevelt Road at Lake Shore Drive, Chicago, IL 60605

D. A. Posey, Laboratório de Etnobiologia, a/c Departamento de Biologia, Universidade Federal do Maranhão, 65.000 São Luís, Maranhão, Brazil

G. T. Prance, The New York Botanical Garden, Bronx, NY 10458

M. E. van den Berg, CNPq/Museu Goeldi, C. P. 603, 66.000 Belém, Pará, Brazil

Foreword

The New York Botanical Garden is pleased to establish the new monographic series ADVANCES IN ECONOMIC BOTANY. This series, the official publication of our Institute of Economic Botany, is intended to be an outlet for the publication of symposia and lengthy research papers in the field of economic botany. With this series we aim to fill a void in the currently available botanical literature and reinforce our commitment to this subject.

We are especially proud that this volume contains the papers presented during the Society for Economic Botany's 1983 symposium entitled "Ethnobotany in the Neotropics." This symposium, organized by Dr. G. T. Prance, brings together two of the Garden's longstanding commitments: collaboration with the Society for Economic Botany and dedication to neotropical exploration and research. Thus it is fitting that, as the Society celebrates its 25th anniversary and ECONOMIC BOTANY nears its 40th anniversary, we join efforts in the dissemination of knowledge on a subject that is of mutual interest.

On behalf of The New York Botanical Garden I thank Dr. Jacquelyn A. Kallunki, Dr. Ghillean T. Prance, and the Society for Economic Botany for their efforts in the preparation, presentation, and publication of this symposium.

We look forward to a series that measures up to the high standards we have set for it, provides a new service to our colleagues, and further strengthens our commitment, begun in 1896, to the publication of scholarly works in botany.

María L. Lebrón-Luteyn
The New York Botanical Garden
7 June 1984

Table of Contents

Introduction

The papers in this first volume of *Advances in Economic Botany* are from a symposium entitled "Ethnobotany in the Neotropics" which sought to bring together presentations representing the current status of ethnobotany in that region. It shows that the subject is very much alive in the area. The symposium covers a variety of research approaches by both residents and non-residents of the Neotropics. However, all authors have considerable field experience in the region. It also covers a wide range of plant groups (from fungi to palms) and examples of many different types of plant uses. To adequately represent contemporary ethnobotany it is significant that three of the authors (Ayala, Posey, and van den Berg) reside in tropical countries and that the authors are a mixture of botanists and anthropologists.

The papers by Berlin, Boster, and Posey represent the anthropological contribution. It is far too frequently the case that anthropological interest in plants and botanical interest in human culture and in uses of plants have been isolated. Yet progress in ethnobotany is made through a blending of methods and information obtained from both fields.

Another significant difference from previous publications on ethnobotany is that this volume is not merely a catalogue of plants used by different indigenous peoples. Much additional information is provided. One of the most important areas of ethnobotany which has recently gained more attention is the greater emphasis on the ecology of indigenous populations. The paper by Posey demonstrates how important it is to study how the people cultivate the plants or manage the forest. Such ecological data are probably more important than lists of plant uses because they produce information which can lead to better use of many underexploited plants with economic potential. Indigenous cultures are not all "shifting cultivators"; some groups have methods remarkably similar to recently emphasized agroforestry techniques.

Heiser and Balick discuss two of the plant families, the Solanaceae and Arecaceae, that have contributed a large number of economic products from the Neotropics. Ayala, Lewis and Elwin-Lewis, and Plowman bring out the richness of the region as a source of medicinal and narcotic plants.

Berlin emphasizes, with some good examples, that native American collectors are becoming an increasingly important and interesting aspect of ethnobotanical research. We can learn much more from them than we can solely from information gathered by strangers from another distant culture. The number of varieties of cultivated plants is often overlooked by collectors who are interested in novelties, but Boster shows that the Aguaruna of Peru can recognize many cultivars of *Manihot*. We need much more data as well as germplasm collections of cultivars of the many different crops utilized by the indigenous peoples of the Neotropics.

We probably have all experienced the marvels of tropical markets. One of the best is the Ver-o-Peso market of Belém, Brazil which is discussed by van den Berg. How important it is for ethnobotany to understand the culture and the economics of the markets that sell a host of little-known, useful plants and that should be a focus of ethnobotany.

I would like to thank both the Society for Economic Botany, the Willard Sherman Turrell Herbarium Fund, Northern Kentucky University, Miami University and the New York Botanical Garden for travel support to enable the speakers to take part in the symposium.

Ghillean T. Prance
The New York Botanical Garden
8 June 1984

Notes on Some Medicinal and Poisonous Plants of Amazonian Peru

Franklin Ayala Flores

Since 1972 I have been conducting ethnobotanical studies in Amazonian Peru especially in the northeastern, northwestern, and southern parts of Loreto Department. My work has concentrated on the Achual, Bora, Candoshi-Shapra, Huitoto, Ocaina, Yagua, and Shipibo tribes which are located along the Tigre, Huasaga, Corrientes, Ampiyacu, Yaguasyacu, Pastaza, Morona, Marañón, Napo, and Ucayali Rivers. The Indians have an extensive knowledge of the uses of plant species for medicine, textile manufacturing, fishing, hunting, home construction, and alimentation.

The Peruvian lowland Amazon Basin is characterized by high precipitation during most of the year, good drainage of the soil, lack of a nutrient-rich layer of organic matter, and an abundance of endemic species. Through time the natives have formed close associations with various plants which form an integral part of their folklore, ritual, and medicine. In this paper I have chosen to concentrate on their uses as medicines and poisons. The family and common names of the genera and species mentioned in the text are given in Table I.

Plants which contain alkaloid hypotensive drugs

Well-known hypotensive alkaloids are obtained from species of *Rauvolfia*, especially *R. serpentina*. This species, which is not native to Peru, contains more than 50 alkaloids of which reserpine and rescinamine are the most important and are used extensively in Western medicine. These drugs have two pharmacologically well-defined actions, a general sedative action which is intense and prolonged and a hypotensive action. Rescinamine, which has a hypotensive action and also acts as a tranquilizer (Font Quer, 1969), is present also in other species of *Rauvolfia*.

In Amazonian Peru, reserpine and other alkaloids such as tetraphylline and tetraphyllicine (Lewis & Lewis, 1977) are obtained from the root of *Rauvolfia tetraphylla* and *R. sprucei*. The Shipibos, Yaguas, and Achuals use these species primarily as an arrow poison, which is made from the roots, rather than as a

Table I

Plant species mentioned in text

Scientific name	Family	Common name (Indian tribe)
Abuta imene (Mart.) Eichl.	Menispermaceae	
A. rufescens Aubl.	Menispermaceae	
Alchornea castaneifolia (Willd.) Juss.	Euphorbiaceae	ipurosa, ipururo, ipururu (Shipibo)
Ambelania lopezii Woodson	Apocynaceae	
Anacardium occidentale L.	Anacardiaceae	cashú, marañón
Anthurium sp.	Araceae	jergón quiro (Achual)
Asplundia sp.	Cyclanthaceae	calzón panga
Brunfelsia grandiflora D. Don subsp. *schultesii* Plowman	Solanaceae	chiric sangango
Calathea sp.	Marantaceae	guisador
Calatola costaricense Standl.	Icacinaceae	piu, pio
Calycophyllum spruceanum (Benth.) Hook.	Rubiaceae	capirona colorada
Campsiandra angustifolia Spr. ex Benth.	Leguminosae	huacapurana
Carica papaya L.	Caricaceae	papaya
Caryocar glabrum (Aubl.) Pers.	Caryocaraceae	almendra
Cassia reticulata Willd.	Leguminosae	retama
Chenopodium ambrosioides L. var. *anthelminticum* (L.) A. Gray	Chenopodiaceae	paico
Chlorophora tinctoria (L.) Gaud.	Moraceae	insira
Chondrodendron limacifolium (Diels) Mold.	Menispermaceae	
Chondrodendron tomentosum Ruíz & Pavón	Menispermaceae	amphihuasca, curare
Clibadium asperum (Aubl.) DC.	Asteraceae	huaca
Couroupita guianensis Aubl.	Lecythidaceae	ayahuma
Crescentia cujete L.	Bignoniaceae	huingo
Curarea toxicofera Barneby & Krukoff	Menispermaceae	
Cyperus sp.	Cyperaceae	vibora piripiri (Achual)
Distictella racemosa Urban	Bignoniaceae	
Eleutherine bulbosa (Mill.) Urban	Iridaceae	yaguar piripiri
Ficus insipida Willd.	Moraceae	oje
Genipa americana L.	Rubiaceae	huito
Gossypium barbadense L.	Malvaceae	algodón, algodón silvestre
Heliconia sp.	Heliconiaceae	bijao, situlli
Hura crepitans L.	Euphorbiaceae	catahua
Inga coriacea (Pers.) Desv.	Leguminosae	bushilla
Iriartea exorrhiza Mart.	Arecaceae	cashapona
Lonchocarpus nicou (Aubl.) DC.	Leguminosae	barbasco, barbasco legítimo, coñapi, cube, huasca barbasco, pacai
Manettia divaricata Wernham	Rubiaceae	yumanasa

Table I

Continued

Scientific name	Family	Common name (Indian tribe)
Mansoa alliacea (Lam.) Gentry	Bignoniaceae	ajo sacha
Martinella obovata (H.B.K.) Bureau & K. Schum. ex Mart.	Bignoniaceae	remio
Maytenus ebenifolia Reiss.	Celastraceae	chuchuhuasi
Mendoncia sp.	Acanthaceae	
Momordica charantia L.	Cucurbitaceae	papailla
Musa paradisiaca L.	Musaceae	banana
Nealchornea sp.	Euphorbiaceae	
Neea parviflora Poepp. & Endl.	Nyctaginaceae	piosha, yanamuco
Paspalum conjugatum Berg	Poaceae	torurco
Phyllanthus acuminatus Vahl	Euphorbiaceae	
Physalis angulata L.	Solanaceae	bolsa mullaca
Piper sp.	Piperaceae	cordoncillo
Pithecolobium laetum (Poepp. & Endl.) Benth.	Leguminosae	remo caspi
Potalia amara Aubl.	Loganiaceae	curarina, sacha curarina
Psidium guajava L.	Myrtaceae	guyaba
Rauvolfia serpentina Benth.	Apocynaceae	
R. sprucei Muell. Arg.	Apocynaceae	huevo de gato
R. tetraphylla L.	Apocynaceae	misho runto
Schoenobiblus peruvianus Standl.	Thymelaeaceae	barbasco caspi
Simira rubescens (Benth.) Bremek. ex Steyerm.	Rubiaceae	pucaquiro caspi
Solanum mammosum L.	Solanaceae	veneno, tintoma, reconilla dulce
Spathicalyx xanthophylla (Mart. ex DC.) A. Gentry	Bignoniaceae	
Spondias mombin L.	Anacardiaceae	ubos
Stachytarpheta cayennensis (L. C. Rich.) Vahl	Verbenaceae	sacha verbena, verbena regional
Telitoxicum minutiflorum (Diels) Mold.	Menispermaceae	
T. peruvianum Mold.	Menispermaceae	
Tephrosia sinapou (Buchot) A. Chev.	Leguminosae	barbasco, cube ordinario, muyuy cube, tingui de cayene, tirana barbasco
Triplaris surinamensis Cham.	Polygonaceae	tangarana

medicine, but Macbride (1959: 376) records the use of *R. tetraphylla* for treatment of nervous disorders.

Plants which contain alkaloid relaxants of the smooth or skeletal muscles

Chondrodendron tomentosum contains the active curare principle D-tubocurarine. This alkaloid, of the bisbenzylquinoline group, is obtained as crystalline chlorohydrate of the alkaloid quaternary base, D-tubocurarine (Font Quer, 1969).

The natives extract it by macerating the bark of the root and stem. They call this bark product "ampi"; hence the local name of the plant is "amphiuasca." They use this extract as arrow poison for hunting and, formerly, in warfare. Although these alkaloids are now used medically to produce muscle relaxation during surgery, particularly abdominal surgery, to avoid the risk of toxic or lethal doses of anaesthetic, the indigenous people of Peru use it only as an arrow poison. In Western medicine, it is also used to decrease the strength of muscle contraction caused by electric shock applied in some types of psychiatric treatment to prevent fractures of the vertebrae (Font Quer, 1969).

Other Peruvian species of Menispermaceae such as *Curarea toxicofera, Telitoxicum peruvianum,* and *Abuta rufescens* are also used in curare preparations. These species are chemically similar to *Chondrodendron limacifolium, Telitoxicum minutiflorum,* and *Abuta imene* which have been reported to be used in various curare arrow poisons (Krukoff & Moldenke, 1938).

The alkaloids curine, isochondodendrine (of the bisbenzylquinoline group) and toxicoferine (which does not resemble any known bisbenzylquinoline in its physical properties) have been isolated from *Curarea toxicofera* (cited as *Chondrodendron* in Cava et al., 1969). From *Telitoxicum peruvianum* the alkaloids norrufescine, lysicamine, and subsesseline as well as telitoxine, peruvianine, and telazoline have been isolated (Menacherry & Cava, 1981). The oxoaporphines—imenine, homomoschatoline, and imerubrine—and the azafluoranthenes—imeluteine, rufescine, and norrufescine—have been found in *Abuta rufescens* (Cava et al., 1975).

The contemporary medicinal use of alkaloids from two plants, *Rauvolfia* and *Chondrodendron,* which are used as arrow poisons by indigenous tribes, demonstrates that we should not just look at indigenous medicinal plants for potential new drugs. Various types of poisons appear to be equally important as sources of useful drugs.

Anti-rheumatic plants

I have selected five species, belonging to five families, which are most commonly used as anti-rheumatics by the natives. The Candoshi-Shapra Indians and the Shipibos use the bark and root of *Alchornea castaneifolia* for the treatment of rheumatism and it is also used by the herbal doctors ("curanderos") of Iquitos who recommend it as the most effective anti-rheumatic. The Achual Indians use *Brunfelsia grandiflora* subsp. *schultesii* and the root of *Mansoa alliacea.* The active principles of the latter species are dimethyl sulfide, divinyl sulfide, diallyl sulfide, propylallyl disulfide, alline, allicine, disulfoxide allyl (López G. & Kiyán de Cornelio, 1974). Its leaves are also used to relieve seasickness and headaches and are effective tranquilizers. Other species used against rheumatism in Amazonian Peru are *Maytenus ebenifolia* and *Campsiandra angustifolia.* The bark of these species is soaked in the local "fire water" ("aguardiente") to make the medicine, which is taken before breakfast.

Plants used against snake bites

The effect, if any, and correct usage of the many plants used to treat snake bites are somewhat controversial. However, four species are widely used among different tribal groups in different places in Loreto Department. The widespread use of these species indicates a real efficacy.

The preferred snake bite remedy of the Bora, Huitoto, Ocaina, and Yagua

Indians is *Potalia amara* which contains squalene and fatty acid methyl esters (Schultes, 1978). The Indians boil the bark of the root to obtain a black substance, which is then filtered. The orally administered dose for a child is five drops and for an adult a teaspoonful (Jose Terres, pers. comm.). The Bora Indians from Brillo Nuevo dig up the root together with the attached mud and then macerate the muddy bark of the root in a small amount of water and place the mixture on the bite (Guillermo Criollo, pers. comm.). The infusion resulting from boiling or soaking the finely chopped leaves in water is drunk "to calm the body and eliminate pain" (Schultes, 1978). The Achual Indians from the Río Tigre employ the corm of *Anthurium* sp. or the rhizome of *Cyperus* sp. as a snakebite treatment. The corm or the rhizome must be pounded and placed immediately on the bite. The Achuals near the Río Corrientes chew about 200 grams of the young petiole of *Asplundia* sp.

Plants used to treat diabetes

In the folk medicine of Iquitos, Peru, the commonest plant used to treat diabetes is *Momordica charantia*. An infusion of the leaves in water taken as a tea before breakfast tends to decrease the blood sugar level for a short time (Elva Reategui, pers. comm.). However, the species which are more often recommended to treat diabetes are *Calycophyllum spruceanum* and *Stachytarpheta cayennensis*. The Indians boil 1 kg of stem bark of the *Calycophyllum* in 10 liters of water until it is reduced to 4 liters. The dose used is 150 ml three times a day for three consecutive months. The stem and leaves of the *Stachytarpheta* are chopped and mixed with a little water. The mixture is then squeezed to obtain a green extract which is taken in doses of one-half a glass once a day for three consecutive months (Benjamin Vasquez, pers. comm.).

The seeds and pericarp of this species contain a saponic glycoside which yields elaterin (a cucurbitacin) and alkaloids that cause vomiting and diarrhea (Lewis & Lewis, 1977). In addition, a pulp of the cotyledons mixed with butter is used to treat infected wounds.

Anti-inflammatory plants

Cassia reticulata is one of the plants most often used as an anti-inflammatory by the Achual Indians and by medicine men ("curanderos") of Iquitos. This species contains antibiotics in the perianth and the leaves as rhein (cassic acid), which is active against gram+ and acid-fast bacteria (Lewis & Lewis, 1977). This species also contains copaiba balsam which is rich in the anthraquinone, emodine, which accounts for its well-known strong, purgative quality (Lewis & Lewis, 1977). The presence of the antibiotic helps to decrease inflammation of the body especially in the case of renal and hepatic diseases. This species is also used to treat venereal and skin diseases.

To treat inflammation of the female genital system the Achuals and the "campesinos" near Iquitos use the fruit and seeds of *Genipa americana* as a genital wash which rapidly heals ulcerated places. The fruit and seeds are boiled and the resulting liquid is used. This species contains genipine, mannitol, tannins, methylester, caffeine, hydantoin, and tannic acid (López G. & Kiyán de Cornelio, 1974). An effective cough medicine is also made by people near Iquitos from the fruit of *Genipa* which has a strong anti-inflammatory effect on the mucous membranes. The Achuals make their cough medicine from the fruit of *Crescentia cujete*. They add 15 drops of vegetable oil and five of kerosene (seven and three, respectively, for children) to the macerated mesocarps of *Crescentia*; this appears to be an effective cure (Benjamin Vasques, pers. comm.).

Plants used to treat hepatitis

Hepatitis is an infectious and contagious disease of the liver, caused by a virus, which can result in serious damage to the hepatic cells and occasionally degenerate into post-hepatic cirrhosis. The Achual Indians use the young root of *Iriartea exorrhiza* as a treatment for hepatitis. They usually pound about 150 g of the root, add about 800 ml of water, and reduce the mixture to one-fourth the volume by boiling. A dose of about 200 ml is drunk before breakfast. The flower of the cotton plant, *Gossypium barbadense* is prepared in a similar way. The root of *Physalis angulata* is drunk as a tea. In the folk medicine of Iquitos, the rhizome of *Calathea* sp. is used more frequently to treat hepatitis. Juice extracted from the rhizome must be taken by the patient in a dose of 15 ml once a day. Another way of preparing the beverage is to pound the rhizome, add it to a glass of lukewarm water, and soak it overnight. The infusion is taken before breakfast.

Plants used in folk odontology

For generations the Candoshi-Shapra Indians from the Morona and Pastaza Rivers have chewed the leaf of *Calatola costaricense*, which stains the teeth black and protects them against cavities. However, the same tribe in Tintiyacu (Andoas, Peru) prefers to chew the leaves of *Neea parviflora* which also stains the teeth black. The same properties are found in a species of *Piper* used by the Yagua Indians from the Itaya River and also by other Indian tribes from Amazonian Peru. The Candoshi-Shapra Indians also chew the seeds of *Manettia divaricata*, which has the same properties as *Calatola*. The Achuals from the Río Tigre chew the bark of *Simira rubescens* to prevent cavities. Chewing the latter dyes the teeth pink or red.

For tooth extraction, the campesinos from the Marañón River use the latex of *Chlorophora tinctoria*. Drops of the latex placed in a tooth cavity cause the tooth to disintegrate. This species also has a diuretic and anti-venereal action, which is due to the presence of moringin, a benzyl amine, which has a great antiseptic value in the urinary tract and is probably a substance that acts against infections of the eyelid (blephorrhagia) (López G. & Kiyán de Cornelio, 1974). The presence in this species of phloroglucine and gallic acid accounts for its antiseptic and astringent properties, respectively.

To relieve toothache, the Indians of the Río Tigre use the leaf bud or young leaf of *Couroupita guianensis*, which is pounded and placed in the cavity. Further details about dental plants of the upper Amazon are given by Lewis and Lewis (1984).

Plants used in ophthalmology

The Candoshi-Shapra from the Río Pastaza use the juice of the root of *Martinella obovata*. They put one or two drops of the juice in the eye to treat conjunctivitis. The natives of Colombia and other parts of Amazonia use the same preparation (A. Gentry, pers. comm.). Bark of *Martinella* with leaves of *Ambelania lopezii* and bark of *Distictella racemosa* are used by the Barasana Indians from Colombia as ingredients of their arrow poison (Schultes, 1970). It is also claimed that an infusion of the bark of *Martinella* is an effective, but dangerous, febrifuge.

The Achuals from the Río Tigre use two drops of the stem sap of *Paspalum conjugatum* to combat conjunctivitis. It is also a good anti-inflammatory. An infusion of the yellow leaves of the liana *Spathicalyx xanthophylla* is used by the Tikuna Indians of Colombia for treatment of severe conjunctivitis (Schultes, 1970).

Anti-malarial plants

In Nuevo Canaan and Intuto on the Río Tigre, the Achuals frequently employ the bark of *Inga coriacea,* a small tree common along river banks, to counteract the effects of malaria. The bark is pounded, steeped in water, strained, and drunk as tea ("agua de tiempo"). The same Indians and some campesinos near Iquitos use in a like manner the bark of *Pithecolobium laetum.* An anti-malarial is also prepared from the bark of *Triplaris surinamensis* by curanderos in the Iquitos area.

Anthelmintic plants

In folk medicine the use of *Chenopodium ambrosioides* var. *anthelminticum* specifically against round worms, *Ascaris lumbricoides* and *Oxyurus vermicularis* ("chicuaca"), is common. It contains ascaridol which acts against intestinal parasites. The white latex of *Ficus insipida* (of which *Ficus anthelmintica* Mart. is a synonym) is used as a vermifuge. It contains phylloxanthine, B-amyrine or lupeol, lavandulol, phyllanthol, and eloxanthine (López G. & Kiyán de Cornelio, 1974). This latter component is probably responsible for the vermifuge action because it is toxic to the parasite. The seeds or latex of the young papaya fruit (*Carica papaya*), which contains papain (an enzyme which aids digestion), is also used as a vermifuge (Arruda Camargo, 1978).

Plants used as anti-diarrhetics

Many different plants employed in the folk medicine of Iquitos possess anti-diarrhetic properties. The wine-red bulb of *Eleutherine bulbosa* is boiled, and the resulting liquid is drunk to treat diarrhea and colic caused either by amoebas, such as *Entamoeba hystolitica,* or bacteria.

For treating diarrhea of bacterial origin, the unripe fruit of *Musa paradisiaca,* the shoot of *Psidium guajava,* the bark of *Anacardium occidentale,* and the bark of *Spondias mombin* are all used. In each case, the plant is boiled and the liquid is drunk. A liquid prepared in the same way from *Spondias mombin* is also used as a vaginal wash for treating infections or hemorrhage.

Poisonous plants

There are many plant species which are used to kill animals. For example, *Solanum mammosum* is a wild plant whose fruit is poisonous and used to kill rats. The Bora Indians from Brillo Nuevo formerly burned the leaves of *Cassia reticulata* to produce a pungent and repellent odor, due to cassic acid, that kills and repels small biting insects, such as *Lutzomyia* sp. (Psychodidae), called "manta blanca." Today this practice is rarely used. The same Indians use the macerated young fruit of this same *Cassia* to treat and eliminate dermatomycosis (Guillermo Criollo, pers. comm.).

Fish poisons are extracted from the root, leaf, fruit, latex, or resin of various plant species. For example, the pounded root of *Lonchocarpus nicou* is mixed with water and sprinkled in rivers, lakes, or streams. The action of rotenone, concentrated in large amounts in the root of the plant, kills the fish in about five minutes. This substance has had great value as a commercial insecticide (Williams, 1936: 215). The most effective fish poisoning process I have seen was at Lagunas, a hamlet located on the Huallaga River in Loreto Department, where *L. nicou* was used in several oxbow lakes. In each case, very many fish were killed and floated to the surface in a short time. Another species used is *Tephrosia sinapou* the roots of which are finely ground and used in the same way as those of *Lonchocarpus nicou*. Another fish poison is prepared from *Schoenobiblus peruvianus*. The triturated root is mixed with water and sprinkled in fishing holes. The cream-colored latex of *Hura crepitans* is very caustic and poisonous and frequently injures the eyes of axmen (Soukup, 1970). The riverside people tap from the tree trunks up to 10 gallons of latex which they use in small lakes and streams to kill fish and anacondas (Jose Torres, pers. comm.). The mesocarp and endocarp of *Caryocar glabrum*, which contains saponins, are pounded and mixed with water and used as a fish poison. The leaf of *Clibadium asperum* is pulverized, mixed with ashes, and used to kill fish. Some natives used to pack this preparation in leaves of *Heliconia* sp. and store it for future use. Other species such as *Phyllanthus acuminatus, Mendoncia* sp., and *Nealchornea* sp. have the same icthyotoxic properties. The active principles in these last three plants have not been identified.

Literature Cited

Arruda Camargo, M. T. 1978. Plantas usadas como anti-helmintico na medicina popular. Ciências y Trópico Recife 6(1): 95.

Cava, P. M., P. J. Kanitomo & A. I. da Rocha. 1969. The alkaloids of *Chondodendron toxicoferum*. Phytochemistry 8: 2341–2343.

———, **K. T. Buck, I. Noguchi, M. Srinivasan & M. G. Rao.** 1975. The alkaloids of *Abuta imene* and *Abuta rufescens*. Tetrahedron 31: 1667–1669.

Font Quer, P. 1969. Medicamenta. Editorial Labor, S. A., Madrid, Spain.

Krukoff, B. A. & H. N. Moldenke. 1938. Studies of American Menispermaceae with special reference to species used in preparation of arrow poison. Brittonia 3: 1–74.

Lewis, W. & M. P. F. Elvin-Lewis. 1977. Medical botany. Plants affecting man's health. John Wiley & Son, New York.

——— & ———. 1984. Plants and dental care among the Jívaro of the Upper Amazon Basin. *In:* G. T. Prance & J. A. Kallunki, editors. Ethnobotany in the Neotropics. Adv. Econ. Bot. 1.

López G., J. E. & I. Kiyán de Cornelio. 1974. Plantas medicinales del Perú. Biota 10: 28–41, 76–84.

Macbride, J. F. 1959. Flora of Peru. Fieldiana, Bot. 13, Part 5, No. 1.

Menacherry, D. M. & P. M. Cava. 1981. The alkaloids of *Telitoxicum peruvianum.* J. Nat. Prod. 44(3): 320.

Schultes, R. E. 1970. De plantis toxicariis e mundo novo tropicale commentationes VI. Notas etnotoxicologicas acerca de la flora Amazonica de Colombia. Pages 178–196. *In:* J. M. Idrobo, editor. II Simposio y Foro de la Biología Tropical Amazonica. Editorial PAX, Bogota, Colombia.

———. 1978. De plantis toxicariis e mundo novo tropicale commentationes XXIII. Bot. Mus. Leafl. 26: 177–197.

Soukup, J. 1970. Vocabulario de los nombres vulgares de la Flora Peruana. Colegio Salesiano, Lima, Peru.

Williams, L. 1936. Woods of northeastern Peru. Field Mus. Nat. Hist., Bot. Ser. 15: 1–587.

Ethnobotany of Palms in the Neotropics

Michael J. Balick

Palms are among the most useful plants in the diverse and complex biological world that is the neotropical forest. For the nomadic hunter-gatherer tribes of the Amazon who depend on the forest for sustenance, palms provide a wealth of products that make their existence in this ecosystem possible. The importance of palms is continually reaffirmed by the lowland farmer who, when clearing away the forest, carefully leaves the palms to be managed and harvested on a semi-permanent basis or utilized as forage and shade by his domestic animals. Even large-scale agribusiness operations find the cultivation of palms to be immensely important and profitable, as shown by the continued increase in African oil palm plantations in many neotropical regions as well as the sustained interest in cultivating the coconut palm. The uses of palms in the Neotropics easily number in the thousands when one considers the variety of ways they are employed in everyday life.

Although they are economically important, from a biological perspective palms are among the least understood elements of the neotropical ecosystem. Despite their long history of observation and study, their striking presence, and their obvious utility, palms are not well represented in scientific collections, and consequently, their taxonomy is less well-known relative to other groups. In a recent paper, Balick et al. (1982) analyzed the collections in herbaria in Brazilian Amazonia and concluded that only 37.5% of the 232 currently recognized palm species in that area were represented in regional collections. Additionally, many specimens lacked some of the most crucial diagnostic elements such as flowers and fruits.

Problems in the study of palms arise from their massive size and weight—up to several metric tons in some species—as well as from the difficulty of collecting them. In order to secure samples of fruits or flowers, it is usually necessary to fell the tree, a task which requires many hours of arduous effort.

History of palm study in the Neotropics

Much of our knowledge of palms has been obtained during several centuries of exploration and study, often under the most difficult conditions. Corner (1966), in his popular work on the natural history of this family, discussed the efforts of those whom he called "palm pioneers."

Advances in Economic Botany 1: 9–23, 1984
© 1984 The New York Botanical Garden

9

Of the neotropical "palm pioneers," Nicholas Jacquin was one of the earliest. He travelled to the Dutch colonies in tropical America and in 1763 described the economically important genus *Bactris.* Hipólito Ruíz and José Pavón explored in Peru and Chile from 1778 to 1788 and, among many other contributions, described the genus *Phytelephas,* the ivory-nut palm.

The well known voyage of Alexander von Humboldt (1799–1804) to the Neotropics was a bridge between two centuries of palm exploration. Humboldt was a naturalist in the truest sense, well versed in many important areas necessary for such study. Along with Aimé Bonpland, Humboldt was sent by the Spanish government to undertake an expedition to its colonies in the New World. Humboldt, impressed by the diversity and splendor of the palms, wrote (1850: 223) that they were ". . . the loftiest and noblest of all vegetable forms . . . ." His writings (1853: 9) are also filled with ethnobotanical observations such as that on the Guaraon Indians' relationship with *Mauritia flexuosa* L.:

> It is curious to observe in the lowest degree of human civilization the existence of a whole tribe depending on one single species of palm tree, similar to those species of insects which feed on one and the same flower . . . .

While contemporary scientists might disagree with his reference to the low degree of Guaraon civilization, the dependence of an indigenous group on one or a few palm species for so many of their daily requirements is a common observation. The Indian use of palms and of other plants in their environment, far from being indicative of a primitive civilization, is actually very complex and dynamic. Not only are great numbers of plants utilized in a complex hierarchy, but that this system is sustainable indicates its advanced nature, having been refined over centuries by the most rigorous experimental trials. Among the contributions of Humboldt, Bonpland, and their colleague Karl Sigismund Kunth, were descriptions of the useful palm genera *Attalea* and *Ceroxylon.*

The person known as the "father of palms" was Carl Friederich Philipp von Martius. From 1817 to 1820 he undertook an expedition to South America that yielded the knowledge to describe a number of economically important palm genera including *Acrocomia, Copernicia, Desmoncus, Guilielma, Leopoldinia, Lepidocaryum, Maximiliana, Oenocarpus,* and *Syagrus.* His major work, completed during his lifetime, was *Historia Naturalis Palmarum,* still one of the finest treatments ever produced of this family.

Hermann Karsten carried out studies in Venezuela and Colombia, making several trips from 1843 to 1852. Although his major research was on the flora of Colombia as a whole, he described the useful palm genera *Jessenia, Scheelea,* and *Socratea.*

During explorations of the Rio Negro and Upper Amazon, Alfred Russell Wallace focused much of his efforts on palms. He lived with Indian tribes and developed a profound understanding of the interdependence of palms and people in the Amazon Valley. This relationship was expressed in a small book, *Palm Trees of the Amazon and their Uses* (1853) which covered 48 species, 14 of which were newly described. Wallace's work has endured as a timeless testimonial to the utility of palms. The vast majority of these uses still can be observed today by travellers to this region.

During his travels in the Amazon, Wallace encountered Richard Spruce, an Englishman who was to spend 15 years in South America from 1849 to 1864 and who has been called by some the greatest botanical explorer of South America. Spruce also lived among the Indians and their palms and developed a remarkable insight into the ethnobotanical utilization and ecology of palms. Spruce offered to collaborate with Wallace in work on the palms, an offer which the latter refused. Spruce published a major paper on the palms (1871), as well as an account of his travels which included many observations on palm ethnobotany, published posthumously in 1908.

The first South American to make substantial contributions to the knowledge of palms and their uses in his own country was João Barbosa Rodrigues. His studies in the Amazon Valley from 1871 to 1874 laid the foundation for his two-volume work, *Sertum Palmarum Brasiliensium* (1903). This massive folio with its glorious, colored plates accurately depicts many of the Brazilian species and describes their regional utilization.

There are a number of other workers who built upon these early studies. Perhaps the greatest contemporary student of the Palmae was Harold E. Moore, Jr. (1917–1980). During several decades of study he logged tens of thousands of miles to all parts of the globe in search of palms and information on their biology and utilization. His reflection on the uses of palms in the Neotropics is as eloquent and accurate a statement as has been made in this regard (Moore, 1973: 64–65):

> Man, however, is the animal in the South American forest that utilizes palms in the greatest number of ways. . . . The versatility of palms in the hands of man is astonishing. Houses, baskets, mats, hammocks, cradles, quivers, packbaskets, impromptu shelters, blowpipes, bows, starch, wine, protein from insect larvae, fruit, beverages, flour, oil, ornaments, loincloths, cassava graters, medicines, magic, perfume—all are derived from palms. The importance of man as a biotic factor in the tropical ecosystem has been argued (Richards, 1952, 1963; C. O. Sauer, 1958). However, to whatever extent man has been involved in the tropical ecosystem, palms have certainly been a major factor in making possible this involvement and even today, despite the advent of the corrugated tin roof and the rifle, they are of primary importance to many primitive American cultures.

The lesson to be learned from these great scientists and their studies, which today form the foundation for all palm research, is that lengthy, in situ field research is absolutely necessary to fully appreciate and understand the powerful and lasting bond between palms and people in the Neotropics.

Palms, their unique nature and diversity of usage

One of the reasons palms are used so ubiquitously throughout their range is that they possess a physical construction differing from other plants, one which provides them with certain advantages for utilization. Within the stem is a series of small strands passing through a matrix of starchy ground tissue. These strands, up to 2–3 mm in diameter, comprise three parts: phloem to conduct nutrients, xylem to conduct water, and fibers that incompletely sheath the strand and offer mechanical support. The strands ring the periphery of the palm stem and account for its strength and flexible nature. Mechanically this is an efficient way to support a heavy object such as a palm. Its durability is proven in the most powerful of tropical storms, when palm stems often survive intact while those of other plant groups are broken and scattered like toothpicks. I have witnessed the steel head of an ax chipping or breaking when used to fell a mature palm specimen, attesting to the strength of the palm stem.

Figure 1 illustrates some of the diverse ways in which palm stem can be employed. Shown at the top is a house constructed with frames of palm stems, from genera such as *Oenocarpus, Jessenia, Mauritia, Syagrus* and others, as well as flooring from strips cut from the stem of *Iriartea* and *Socratea* species. Moving clockwise, are illustrated an arrow point with reverse barbs fashioned from the fibrous bundles of the stem of *Jessenia bataua* (Mart.) Burret and a bow from

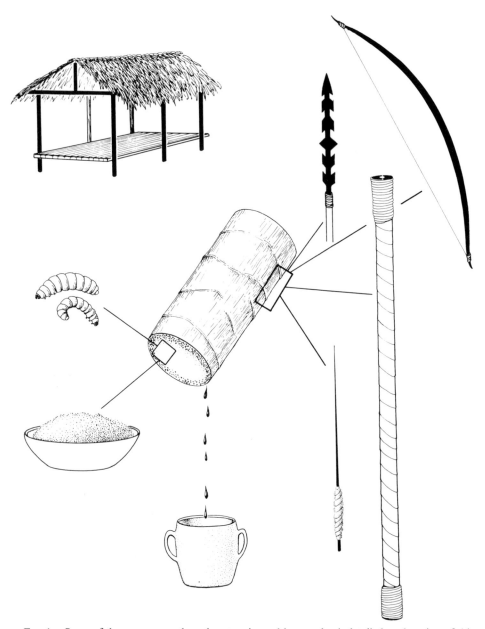

Fig. 1. Some of the many ways the palm stem is used by people. A detailed explanation of this plate can be found in the text. Drawn by Bobbi Angell.

wood of the same species. The stem of *Bactris gasipaes* H.B.K. is split in two, grooved on the inside of each half and bound together to make a blowgun. A one-piece blowgun is made from the slender stem of *Iriartella setigera* (Mart.) H. Wendl. which is hollowed by removing the ground tissue with a long stick. The petiolar spines of *Jessenia bataua* are used to make the darts for blowguns. In addition, thread from *Astrocaryum* species is used to affix to the base of the dart a piece of kapok which forms an airtight seal in the blowgun to propel the dart. The stems of *Orbignya* species are cut and the sugary liquid that freely flows out is collected and fermented for consumption. The starch stored in the ground tissue of *Manicaria saccifera* Gaertn. and *Mauritia flexuosa* is collected and consumed as an important carbohydrate food; sago is collected in a similar manner from palms in the Old World (Ruddle et al., 1978). When palms are cut, for whatever reason, a section of the cut stem is often left lying on the forest floor. Within minutes of cutting, beetles begin flying around this stem and bore into it to lay their eggs. The larvae develop rapidly and within a month or two, when they are a few inches long, the people harvest the insects which are an excellent source of protein. Such larvae, e.g., that of the weevil *Rhynchophorus palmarum,* are consumed raw, fried, or boiled.

This same diversity of usage can be found for the palm leaf, due to its unique construction, variation of form, and size. Both palmate and pinnate leaves are represented in Figure 2. Beginning from the top and moving clockwise, the first diagram depicts the use of a *Manicaria saccifera* leaf as a sail for river transport along the Orinoco (Wilbert, 1980). Many kinds of palm leaves are used as thatch for houses and other structures. The Bora Indians of Peru use the seedlings of *Jessenia bataua* to treat snakebite. Leaves of *Geonoma* species are burned and used to produce a salt substitute in some areas of the Amazon. Evil spirits are believed to assume the form of some palm species, such as *Lepidocaryum tenue* Mart., and to terrorize local inhabitants. The palmito (palm heart) from *Euterpe* species as well as from other genera forms the basis for a substantial commercial export industry in both Central and South America. Filters for various indigenous preparations are made from rolled leaves of *Geonoma* species, which also serve as drinking cups. Wax from *Copernicia prunifera* (Miller) H. E. Moore is used as a medicinal plaster for treatment of wounds and other skin injuries. One of the most durable plant fibers is derived from the leaf of *Astrocaryum vulgare* Mart., which is then woven into hammocks, bags, nets and other useful items. Finally, the foam-like petioles of *Mauritia flexuosa* are bound together to make rafts for fishing.

Because palms have adapted to a great variety of habitats they are relatively abundant in neotropical ecosystems and thus available for widespread use. In whatever tropical lowland location Indians choose to inhabit—the seasonally inundated floodplains for cultivating its rich, fertile soil, the savannah or virgin tropical forest for hunting, the river bank for fishing, or the gallery forest for hunting as well as agriculture—palms are usually present. Survival of palms in inundated habitats is enhanced through specialized structures known as pneumatodes which allow for gaseous exchange in wet areas. De Granville (1974) distinguished two types of these structures, those which appear as rings on the root system and those which appear as small conical protrusions from the roots. He concluded that these types of pneumatodes could serve as stable taxonomic characters to distinguish palm species. Species such as *Mauritia flexuosa* and *Euterpe oleracea* Mart. which possess pneumatodes are able to colonize inundated environments and establish pure stands therein.

There are also specialized structures which people employ directly to their advantage. The waxy layer on the leaves of *Copernicia prunifera,* a palm which inhabits hot, exposed habitats, is the basis of a multimillion-dollar industry in Brazil involving harvest and sale of this product, known as carnauba wax in the international trade. The wax of this palm is extremely hard and durable, of high quality, and in great demand (Johnson, 1972). Many different forms of palm

Fig. 2.   Some of the many ways that palm leaves are used by people. A detailed explanation of this plate is found in the text. Drawn by Bobbi Angell.

spines and fibers provide protection against predators such as birds, rodents, and other mammals (Uhl & Moore, 1973). People collect these materials and turn them into rope, darts, arrow points, combs, brooms, and similar products. Fruits with oily outer coatings attractive to dispersal agents are also exploited by people for protein and oil. In northern South America and Trinidad, the fruit of the *Jessenia* palm is a principal food of the guacharo bird (*Steatornis caripensis*) which collects the fruit in the lowlands and deposits it in caves at altitudes of up to 8,000 feet (Ingram, 1958). People also employ the fruits of this palm to produce a protein-rich beverage and a high-quality edible oil.

Generic review

This final section will consider the range of uses for selected palm genera in the Neotropics. These are gathered from the literature and from personal fieldwork with a number of indigenous groups, primarily the Bora Indians of Peru and Guahibo Indians of Colombia.

BACTRIS

Bactris is one of the largest and most widespread neotropical palm genera, perhaps comprising some 200 species (Wessels Boer, 1965). To the taxonomist it is also one of the most confusing and poorly known. It is extremely variable in size and form but usually has spines on the leaves and trunk.

Bactris gasipaes is an important food palm in the Neotropics. Known as "pupunha" in Brazil, "pejibaye" in Costa Rica, "chontaduro" in Colombia, "pijuayo" in Peru, "pichiguao" in Venezuela, and peach palm in English, it is always found as a cultivated plant. When found in the forest or along a river bank at a seemingly unoccupied site it is indicative of prior human occupation. The fruits are somewhat flattened-ovoid and borne in large panicles. When ripe, the fruits are harvested, boiled, and eaten, tasting somewhat like roasted chestnuts. They are considered a national food in Costa Rica, commonly sold along the streets and eaten by poor and rich alike. Their nutritional composition per 100 grams edible portion is 196 calories, 2.6 g protein, 4.4 g fat, 41.7 g total carbohydrate, 1.0 g fiber, and 0.8 g ash. The peach palm is rich in potassium (46 mg/100 g), Vitamin A (670 mcg/100 g), riboflavin (0.16 mg/100 g) and Vitamin C (35 mg/100 g) (Leung, 1961).

The heart of this species is an excellent substitute for the palmito from wild *Euterpe* species. Because most of the once-vast native stands of *Euterpe* in Costa Rica have been destroyed, the peach palm is now beginning to be cultivated for commercial production of palm hearts in that country. This species grows quickly from seed, suckers when cut, and provides a much larger heart than *Euterpe*. The major disadvantage of the peach palm is the spiny nature of the stem, although some varieties have been selected for the absence of these fierce spines. Work is underway at several regional research centers in Costa Rica and Brazil to select and further develop this palm for wider commercial use.

Among the other species of *Bactris, B. macroacantha* Mart. is used by the Bora Indians of Peru as a soporific. The round yellow fruits of this diminutive species are edible, quite sweet, and, according to local belief, make a person relaxed or sleepy, depending on how much fruit is consumed.

The Bora have a legend involving another species, *Bactris fissifrons* Mart., and the creation of the toucan. Usually when young Bora women have their first menstruation they are advised to eat a number of specific foods, but a young woman named Nulleh insisted on eating only the tender shoots of this palm. The spiny leaf stuck to her tongue and she could not remove it. She then turned into a toucan and flew into the forest, the leaf becoming her beak and her long hair becoming feathery plumage. When a person dies, the body lice are said to leave the corpse and go to the toucan, who then realizes a person is dead and cries for them. Thus, the toucan is an important bird to the Bora, its cry signifying death.

EUTERPE

Euterpe is a genus of graceful palms up to 20 meters tall, with slender, solitary or caespitose trunks. It is widely distributed throughout the Neotropics in a number of different habitats but is especially abundant in swampy or moist areas. Glassman (1972) lists 49 species in his checklist of American palms. There is great need for a better understanding of the taxonomy of this group; it probably is comprised of far fewer species.

The most important commercial use of this genus is as a source of palmito or heart of palm. Vast natural groves of *Euterpe oleracea* in the Amazon are felled to extract the young growing tip and developing leaves in the crownshaft. Each tree produces only a few segments of heart for canning and when harvested for this purpose, the rest of the tree is abandoned and wasted. While working in an extremely remote region of the Amazon Valley, I encountered a team of men who were systematically mapping large zones of *Euterpe* in the forest for commercial exploitation. When their inventory was completed, portable canning factories were to be brought in and the stands decimated in a way reminiscent of the manner in which the buffalo were hunted down only for hides on the North American frontier.

The exploitation of palmito has become a major industry in the Neotropics, and in Brazil alone some 114,408 tons were harvested in 1980, primarily from *Euterpe* species (IBGE, 1982).

The beverage known as "assai" in Brazilian Amazonia is produced from the fruit of *Euterpe oleracea*. The fruit is formed in large panicles, each weighing several kilos. When ripe, the fruit turns a deep purple and the trunk must be climbed to carefully remove the panicles. The oil-rich beverage, produced from the mesocarp, is sold at small roadside stands. It may be mixed with farinha or occasionally with sugar. In 1980, about 60,000 tons of assai fruit entered local commerce in Brazil, mostly from the State of Pará (IBGE, 1982). Nutritional studies mentioned by Cavalcante (1977) indicate that assai has more calories per unit than milk and twice as much fat. During one trip to the interior, I was cautioned to avoid the following "harmful combinations": assai juice consumed with alcohol, with mango, or with juice of *Oenocarpus bacaba* Mart. Eating assai with any of these foods was said to have deleterious effects ranging from stomach pain to serious illness. On another occasion I was told to eat a popsicle of assai to calm an upset stomach, which turned out to be quite effective.

Goulding (1980) in his excellent study of the relationship between the Amazon forest and fish reported that the electric eel (*Electrophorus electricus*) considers *Euterpe oleracea* to be a favorite food. The eels congregate at the base of these palms along riverbanks and in inundated areas where local people believe that the eels shock the trees to knock ripe assai fruits into the water. He could not confirm this supposition, but, correct or not, people refuse to climb *Euterpe* palms in areas where eels are known to be present for fear of being killed.

JESSENIA

The genus *Jessenia* as recognized by Balick (1980) consists of a single species, *J. bataua,* which is further divided into two subspecies. The palms in this genus are large, to 25 meters tall, with massive trunks and a solitary habit. They are found in both inundated and upland regions where they often achieve predominance in an ecosystem. Known as "pataua," "seje," "ungurahui" or "milpesos," the fruit is harvested for human consumption throughout its range in the northern half of South America up to an elevation of about 3500 feet. The fruit is purplish-black with a firm epicarp under which there is a soft, pulpy mesocarp (Fig. 3).

Figs. 3–4. 3. A fruit of *Jessenia bataua* with the epicarp removed to reveal the oily mesocarp. 4. Fruit of *Mauritia flexuosa* purchased in the market in Iquitos, Peru. These fruits are more elongate than those of other species in the Amazon Valley.

The mesocarp contains a high percentage of oil, up to 50% in some cases (National Academy of Sciences, 1975). This oil is physically and chemically almost identical to olive oil. Its major fatty acid components are oleic (77.7%), palmitic (13.2%), stearic (3.2%), and linoleic (2.7%) (Balick & Gershoff, 1981).

Indians harvest the fruits when they are ripe but not yet soft to the touch. The fruits are then soaked in warm water for a few hours or overnight to loosen the epicarp. The water-fruit mixture is macerated and finally filtered to remove bits of epicarp and fibers occurring in the mesocarp. This milky beverage is then consumed alone or mixed with farinha from cassava. The biological value of the protein found in *Jessenia* fruit is extremely high, comparable to that of good animal proteins and much better than most grain and legume proteins, making this beverage an excellent source of nutrition.

As previously mentioned, the petiolar spines serve as blowgun darts (when wrapped with kapok and tipped with curare arrow poison), and wood from the stem is employed as arrowpoints and for bows. The Bora Indians of Peru use the leaves of this species to construct provisional pack baskets for hunting. When an animal is killed a basket is woven in a few minutes to carry the meat back to the village. Leaves also serve as thatch for houses and for walls, room dividers, and chicken coops.

The oil expressed from the mesocarp forms the basis for a cottage industry wherever the palms are found in quantity. In addition to its use as an edible or cooking oil, the Guahibo Indians of Colombia and Venezuela use the oil as a remedy for tuberculosis, cough, asthma, and other respiratory problems and as a hair tonic.

Lepidocaryum

The genus *Lepidocaryum* is represented by only a handful of species in the Neotropics of which *L. tenue* and *L. gracile* Mart. are the most common. These are diminutive understory palms which grow in primary or disturbed forest. In the areas where these palms are found, they are considered the finest thatch for dwellings and other structures, lasting many years without need of replacement or patching. The Bora Indians of Peru use *L. gracile* as a remedy for ocular infections. The thin stem, a few centimeters in diameter, is roasted until soft and the juice contained within squeezed into the eye. The curative properties are said to be similar to the effects of antibiotics.

Because some Indians of the Northwest Amazon believe that a huge monster, the "curipira," often becomes small enough to hide in the groves of these plants and takes on the form of the palm itself to terrorize and harm them, they are sometimes reluctant to gather *Lepidocaryum tenue,* even though it may be abundant and provide the best thatch (Schultes, 1974).

Mauritia

Mauritia is a widespread genus of some 17 species native to tropical South America and Trinidad. It comprises a number of massive-trunked palms as well as a group having smaller stems that are often covered with woody thorns. One of the most useful species is *M. flexuosa,* known as "moriche" in Venezuela, "aguaje" in Peru, and "muriti" in Brazil. It is a huge palm, to 25 meters or more, and often grows in swampy or moist areas. Its costapalmate leaves are up to 4 meters long. The segments are removed from the petiole, split into several sections and used for thatch by hooking over roof crossbars. This thatch will last from 2–3 years and is, thus, of intermediate durability when compared to other palms used for thatch.

The fruits are relished wherever the palm is found. They are round or ovoid in shape, and underneath the scaly epicarp is found a yellow or orange flesh which contains 12% oil (Balick, 1979). The flesh is peeled from the stony seed and mixed with water to make a beverage or used in confections and ices. The Guahibo strain the juice from the pulpy mesocarp and allow it to ferment for 3–4 days to produce an alcoholic beverage for drinking at festivals or at night after a day's labor.

Fruits of *Mauritia flexuosa* (Fig. 4) are commonly sold in the regional markets where several different forms can be distinguished, based on fruit size, shape, and color. Because the biology of this genus is not well understood, it is not clear whether these forms have taxonomic status.

Another product of this palm which enters the regional economy is the pith from the petioles. The petioles are tubular and two meters or more in length. The foam-like pith floats and is used to cork bottles and make rafts for fishing as well as childrens' toys. While living amongst the Bora we were offered the use of a mattress made of split petiole segments tied together with cord, which in fact was quite comfortable. These petioles are also sold in the Ver-o-Peso market in Belém, Brazil. One of the more curious devices I have ever seen constructed from the petioles of this species looks like triangular crib for young children (Fig. 5). Actually the children stand along the edge of this frame and learn to walk, holding on to its edge as they meander around the center. This was said to be a common device among the Guahibo of Colombia.

This species also provides a useful fiber for weaving. Guahibo hammocks commonly have ornamentation woven from *Mauritia flexuosa* affixed to the sides. A high quality fiber is obtained from the leaves. According to Schultes (1977), the fiber is threadlike and white, and that of the younger leaves is stronger than that of the older leaves. In 1980, Brazil produced 614 tons of fiber from this palm, mostly in the State of Maranhão as well as a small amount in Pará (IBGE, 1982).

MAXIMILIANA

Maximiliana consists of a single species, *M. maripa* (Corr. Serr.) Drude, a tall, solitary palm to 18 meters in height (Glassman, 1978a, 1978b). It is common in northern South America and Trinidad in both well-drained and wet sites. Local names for this species include "inajá" and "cucurito." The leaves, to 8 meters in length, are an excellent source of thatch. The pinnae are folded over to one side and the leaves laid on roof crossbars. The newly emerging leaves are used to weave mats, pack baskets, and walls to divide space in the characteristic, large, open houses of indigenous inhabitants of the Orinoco and Amazon valleys.

The ripe fruit of this species is an excellent food. The kernel contains 60–67% fat and the mesocarp 42.1% fat (Eckey, 1954). In addition to local consumption, sale of the fruit has a small economic impact, appearing in markets in some lowland cities and villages throughout the range of this palm. During a stay in a Cubeo Indian village along the Rio Vaupés in Colombia, I observed the children collecting these fruits, baking them over a fire, cracking them open and consuming the oily kernels as a playtime activity. People commented that this was a common pastime of the children in this particular village, one which provided them with substantial nutritional benefit.

The Guahibo use the endocarp of this species to cap the ends of a Y-shaped device known as a "silípu" or "sirípo" used to snuff "yopo," an hallucinogenic snuff made from *Anadenanthera peregrina* (L.) Spegazzini, a leguminous tree. This snuff tube is placed against the nostrils and the round seeds form a tight fit for inhaling the drug from a small wooden plate.

ORBIGNYA

Orbignya is a widespread genus occurring from Mexico to Bolivia, primarily at low elevations, although it also occurs in areas up to several thousand feet. A preliminary study by Glassman (1977, 1978b) suggested that some 29 species of

FIGS. 5–6.   5. A frame constructed from the petioles of *Mauritia flexuosa* in which Guahibo children support themselves while standing and eventually learn to walk. 6. Grinding the kernels of *Orbignya martiana* by hand in a wooden trough to express the oil they contain.

this genus were described of which 17 "definitely or most probably belong to *Orbignya*." This genus is closely related to a number of other genera including *Attalea, Parascheelea, Maximiliana,* and *Scheelea.* Wessels Boer (1965) lumped these genera together under *Attalea.*

There are several economically important species in the genus *Orbignya.* One of the most important, *O. martiana* B.R., is known as "babassu" in Brazil and "cusi" in Bolivia. An edible oil and a charcoal used in cooking and industry are obtained from the fruit (Fig. 6). In 1980, 250,951 tons of babassu kernels were produced in Brazil, all of which were collected from wild plants. The commercial value of this plant exceeds 60 million dollars annually, making it the largest oilseed industry in the world completely dependent on a wild source. The fatty acid composition of this oil is primarily lauric (44–46%), myristic (15–20%), palmitic (6–9%), caprylic (4.0–6.5%), and stearic (3–6%) (Eckey, 1954). At the present time there is great interest in domesticating this plant, and the Brazilian government, with the assistance of international development agencies and research organizations, is engaged in an effort to collect and document the great variation in these wild palms. A germplasm bank of these trees comprising material collected from several different countries as well as from various regions of Brazil has been developed in Bacabal, Brazil, to help accomplish the goal of domesticating this genus.

One of the local uses for cusi oil in lowland Bolivia is as a remedy for cough. A few drops of liquid from boiled guava leaves (*Psidium guajava* L.) are mixed with a tablespoon of this palm oil and taken four times daily to quell cough. The oil is also massaged directly onto the head for headaches and other head pains and applied once or twice daily to control dandruff. The oily kernels are burned in a flame and rubbed on the eyebrows and on other facial hair to make it darker as well as, according to local belief, to increase its rate of growth. A few teaspoons of cusi oil are said to help an ailing liver. In Brazil the oil of babassu is mixed with sugar and used by some people as a vermifuge.

Charcoal is produced from this species by burning the residue of the fruit after the oily kernels have been extracted. Pits are dug in the ground and filled with the husks, which are then covered with leaves and soil and ignited. After a slow burning process the charcoal is ready for local use in cooking or for sale to industry as fuel. In Bolivia I observed the petioles being burned in bread ovens. When questioned about this practice, people said they preferred this fuel because it burned evenly and cleanly for a long period, allowing the bread and other items to be baked to perfection. In this same area trees are commonly felled to obtain leaves for thatch. About 500 leaves comprise the roof of a small home, measuring some 20 feet long by 10 feet wide, and 50 mature trees are felled to thatch this single dwelling.

This paper has emphasized the beneficial aspects of the palm-people relationship through examples from a very few species. There are also detrimental aspects of this interaction, one of which is illustrated by the results of a study of Chagas' disease done by Whitlaw and Chaniotis (1979). Chagas' disease is an affliction of the poor and of people in areas receiving minimal medical attention, and it is a major problem in Latin America. Although there is no accurate estimate of the total number of people affected, as many as eight million people in Brazil may suffer from this disease (Whitlaw & Chaniotis, 1979). These authors, working in Panama, identified a close relationship between the insect vector of this disease and the corozo palm, *Scheelea zonensis* Bailey. Of 92 randomly examined corozo palms in the Canal Zone, all were found to harbor triatomines, the insect vectors.

The presence of a corozo palm in a particular area was positively correlated with high incidences of Chagas' disease in the human population living in the same area. Conversely, low rates of this disease were found in areas with low densities of corozo palms. Other species of palm trees were found not to harbor these vectors.

The corozo palm is also the home of the principal animal reservoirs of this disease: opossum, anteater, and spiny rat. It is not known why these insects and animals appear in such great abundance on this palm species. However, it would seem advisable for public health workers to begin a widespread study of this disease in order to understand how palms may contribute to the persistence of Chagas' disease. To control the disease, it may be necessary to eliminate breeding sites for the disease vector or to limit the human populations in areas with high concentrations of the palms. This scientific study does give credence to some of the indigenous beliefs in monsters and evil spirits inhabiting palms. It helps to explain why some people consciously avoid dwelling or even walking near these species, many of which are covered with fibers and sheaths—ideal breeding sites for detrimental insects and other harmful organisms.

I have mentioned only a few of the thousands of interactions, both beneficial and detrimental, between palms and people in the Neotropics. We should not lose sight of the fact that palms play an important role in the lives of subsistence peoples as well as those who depend on a cash economy. In 1979, Brazil reported over US $100,000,000 of commerce resulting from the harvest and sale of products from six native palm genera: *Astrocaryum, Attalea, Copernicia, Euterpe, Mauritia,* and *Orbignya* (IBGE, 1981). Much of this economic return was realized in the poorest areas of the country and often represented a significant portion of the cash income of the persons involved. Continuing investigation into this subject will uncover a wealth of new information and provide additional alternatives for land utilization in the tropics.

Literature Cited

Balick, M. J. 1979. Amazonian oil palms of promise: A survey. Econ. Bot. 33: 11–28.

———. 1980. The biology and economics of the *Oenocarpus-Jessenia* (Palmae) complex. Ph.D. dissertation, Harvard University, Cambridge, Massachusetts.

——— **& S. N. Gershoff.** 1981. Nutritional evaluation of the *Jessenia bataua* palm: Source of high quality protein and oil from tropical America. Econ. Bot. 35: 261–271.

———, **A. B. Anderson & M. F. da Silva.** 1982. Palm taxonomy in Brazilian Amazônia: The state of systematic collections in regional herbaria. Brittonia 34: 463–477.

Barbosa Rodrigues, J. 1903. Sertum palmarum Brasiliensium. 2 vols. Bruxelles.

Cavalcante, P. B. 1977. Edible palm fruits of the Brazilian Amazon. Principes 21: 91–102.

Corner, E. J. H. 1966. The natural history of palms. Weidenfeld and Nicolson, London.

De Granville, J. J. 1974. Aperçu sur la structure des pneumatophores de sols hydromorphes en Guyane. Cah. ORSTOM, Sér. Biol. 23: 3–22.

Eckey, E. W. 1954. Vegetable fats and oils. Reinhold, New York.

Glassman, S. F. 1972. A revision of B. E. Dahlgren's index of American palms. J. Cramer, Lehre.

———. 1977. Preliminary taxonomic studies in the palm genus *Orbignya* Mart. Phytologia 36: 89–115.

———. 1978a. Preliminary taxonomic studies in the palm genus *Maximiliana* Mart. Phytologia 38: 161–172.

————. 1978b. Corrections and changes in recent palm articles published in Phytologia. Phytologia 40: 313–315.

Goulding, M. 1980. The fishes and the forest. University of California Press, Berkeley and Los Angeles.

Humboldt, A. von. 1850. Views of nature. Henry G. Bohn, London.

————. 1853. Personal narrative of travels to the equinoctial regions of America. Henry G. Bohn, London.

IBGE. 1981. Produção extrativa vegetal—1979. Vol. 7. Fundação Instituto Brasileiro de Geografia e Estatística-IBGE, Rio de Janeiro.

————. 1982. Produção extrativa vegetal—1980. Vol. 8. Fundação Instituto Brasileiro de Geografia e Estatística-IBGE, Rio de Janeiro.

Ingram, C. 1958. Notes on the habits and structure of the Guacharo *Steatornis caripensis.* Ibis 100: 113–119.

Johnson, D. 1972. The carnauba wax palm (*Copernicia prunifera*). IV. Economic uses. Principes 16: 128–131.

Leung, W. T. W. 1961. Food composition table for use in Latin America. U.S. Gov. Printing Office, Washington, D.C.

Martius, C. F. P. von. 1823–50. Historia naturalis palmarum. 3 vols. Munich.

Moore, H. E. 1973. Palms in the tropical forest ecosystems of Africa and South America. Pages 63–88. *In:* B. J. Meggers, E. S. Ayensu & D. Duckworth, editors. Tropical forest ecosystems in Africa and South America: A comparative review. Smithsonian Institution Press, Washington, D.C.

National Academy of Sciences. 1975. Underexploited tropical plants with promising economic value. National Academy of Sciences, Washington, D.C.

Richards, P. W. 1952. The tropical rainforest. Cambridge University Press, Cambridge.

————. 1963. What the tropics can contribute to ecology. J. Ecol. 51: 231–241.

Ruddle, K., D. Johnson, P. K. Townsend & J. D. Rees. 1978. Palm sago a tropical starch from marginal lands. University Press of Hawaii, Honolulu.

Sauer, C. O. 1958. Man in the ecology of tropical America. Proc. Ninth Pacific Sci. Congress 20: 104–110.

Schultes, R. E. 1974. Palms and religion in the northwest Amazon. Principes 18: 3–21.

————. 1977. Promising structural fiber palms of the Colombian Amazon. Principes 21: 72–82.

Spruce, R. 1871. Palmae Amazonicae, sive enumeratio palmarum in itinere suo per regiones Americae a equatoriales lectarum. J. Linn. Soc., Bot. 11: 65–183.

————. 1908. Notes of a botanist on the Amazon and Andes (A. R. Wallace, editor). Macmillan, London.

Uhl, N. W. & H. E. Moore. 1973. The protection of pollen and ovules in palms. Principes 17: 111–149.

Wallace, A. R. 1853. Palm trees of the Amazon and their uses. John van Voorst, London.

Wessels Boer, J. G. 1965. The indigenous palms of Suriname. E. J. Brill, Leiden.

Whitlaw, J. T. & B. N. Chaniotis. 1979. Palm trees and Chagas' disease in Panama. Am. J. Trop. Med. Hyg. 27: 873–881.

Wilbert, J. 1980. The palm-leaf sail of the Warao Indians. Principes 24: 162–169.

Contributions of Native American Collectors to the Ethnobotany of the Neotropics

Brent Berlin

In an important and highly useful monograph, *Plant Collecting for Anthropologists, Geographers, and Ecologists in New Guinea,* by the well known Australian botanist J. S. Wormersley (1969: 33), one finds a table called "Don'ts in Plant Collecting and Handling." This list of twenty cautionary notes serves as an important reminder for field anthropologists who might see the collaborative role of the botanical specialist as simply that of a technician providing a useful service. Nonetheless, the last "don't" on Wormersley's list suggests a kind of botanical ethnocentrism that is unfortunate. Wormersley states, "Don't send the informant to make the collections" (1969: 34).

If Dr. Wormersley's advice appears ethnocentric, it is, more importantly, simply bad advice. I argue in this paper that not only can one "send the informant to make the collections," but it behooves us as field-workers to develop a truly collaborative effort with native assistants in many other aspects of botanical field work. This collaboration, if elaborated fully, can have enormous payoffs for both the botanist and ethnobotanist alike.

Flora Malesiana describes scores of native collectors who have assisted botanists working in the Old World tropics (Van Steenis-Kruseman, 1974). Comparable efforts in the New World are considerably rarer, although it is not clear why this should be so. A notable exception is Cyrus Lundell who employed the Mayan Indian, Elias Contreras, as a major collector on the Petén of Yucatán. Contereras prepared over 12,000 numbers in sets of seven over a ten-year period in the first half of this century. My longtime collaborator, Dennis E. Breedlove, however, has used and continues to use native collectors as a major component of his botanical field research in Chiapas.

The contributions of native collectors to the ethnobotany of the neotropics that I wish to describe are taken from my own fieldwork experiences with two groups of native Americans, the Tzeltal Maya of Mexico's most southern state of Chiapas—

work carried out in the sixties—and of the Aguaruna and Huambisa Jívaro of the rainforests of north central Peru—work carried out during the seventies (a portion of which is discussed in the paper by Boster, 1984). While a number of Indians in each of these groups participated actively in these ethnobotanical studies, I will focus specifically on the contributions of those key collaborators who worked full-time as ethnobotanical assistants—the Tzeltal Alonso Méndez Ton, the Aguaruna Ernesto Ancuash and Rubio Kayap, and the Huambisa Victor Huashikat and Santiago Tunqui. Alonso Ton is a member of the Tzeltal community of Tenejapa, some twenty miles from the colonial town of San Cristobal de Las Casas. Ernesto Ancuash and Rubio Kayap reside in the community of Huampami on the Cenepa River, Amazonas, Peru. Victor Huashikat and Santiago Tunqui are Huambisa from the village of Caterpiza, a confluent of the Santiago River in Amazonas, Peru.

Ethnobotanical research among the Tzeltal

The Tzeltal plant research began when I serendipitously met Dennis E. Breedlove during the early stages of both of our graduate careers at Stanford University. When the opportunity for starting long-term ethnobotanical fieldwork presented itself, Dennis became the project's field botanist. Earlier I had worked closely with Alonso Méndez Ton on aspects of Tzeltal ethnolinguistics (Berlin, 1962, 1968), and I asked him to continue as an assistant with the new work on ethnobotany. Breedlove began to instruct both Alonso and myself in the procedures of field botany. After a few short months, we asked the young Mayan Indian to set out on his own as an independent collector. We were motivated by two primary reasons. First, our efforts to gain complete coverage of the local flora would be enhanced by adding another full-time collector. Secondly, and as importantly, Alonso would continue to collect on a year-long basis in areas not easily accessible to the gringo participants on the project. Equipped with adequate supplies, Alonso was able to collect, press, dry, and ship plant materials as systematically as a university-trained botanist.

The small initial investment in training Alonso Ton was rewarded by the production of a goldmine of botanical and ethnobotanical information. This point is illustrated by examining the structure of Alonso's collection notebook (Fig. 1). Each entry is comprised of his collection number, his name for the particular specimen, followed by an abbreviated but concise description of the plant's morphology. For example, the first line entered for collection *2307* reads, "b. borbos 'genuine borbos' [*Govenia superba*], one meter in height [are] its stem and leaves, its flowers are yellow."

The entries that appear immediately under Alonso's own name and description are those names provided by other Mayan informants who are present at the time the collection was made. While informants 31, 32, and 50 all apply the same name as Alonso's to this specimen (indicated by the "ditto" marks), there is considerable variation in the names given to other collections, such as for *2308* in Figure 1. Alonso systematically transcribes these variations for each informant present. An analysis of these lexical variants for plant species allows the study of a number of interesting social and cultural variables. For example, distribution of the use of particular names can be mapped, thus providing an index to patterns of social interaction within and between ethnic groups. In addition, the Tzeltal data shows that plants which are valued highly (such as cultivated species) are referred to by significantly fewer synonyms than are plants that have little or no cultural importance.

Alonso has collected more than 4500 specimens on his own and has assisted

147,215

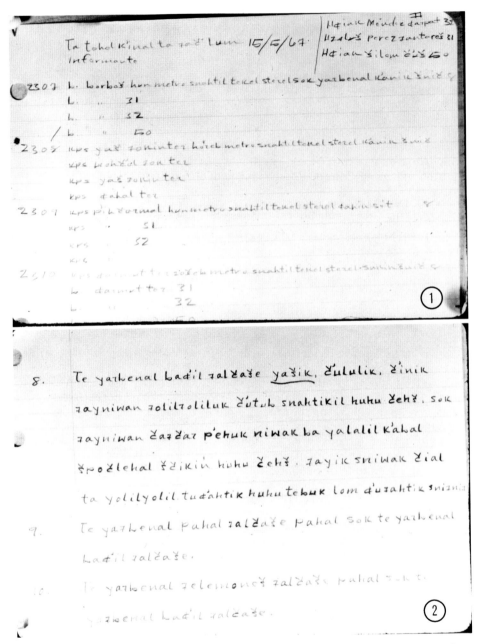

FIGS. 1–2. 1. A page from the collection notebook of Alonso Méndez Ton. 2. A page of text produced by Alonso Méndez Ton that describes the appearance of the leaves of the orange tree (*Citrus* sp.).

Breedlove in the collection of another 15,000 numbers in Breedlove's mammoth *Flora of Chiapas* project. More than 20 new species have been discovered and described as a result of Alonso Ton's collecting efforts, and at least seven have been named in his honor, e.g., *Ilex tonii* Lundell.

The remarkable quality and number of botanical collections made by native assistants is but a single facet of their contributions to ethnobotanical research. Native botanical collectors are particularly important as the source of the cultural information that is essential to all good ethnobotanical studies. With minimal instruction, they are able to provide data on their own system of folk classification of plants as well as important details on the cultural significance of plants in agriculture, ethnomedicine, house construction, the production of tools and weapons, and many additional aspects of ethnobotanical knowledge that bear on the economic importance of plants to human societies.

An example of the kinds of information that can be elicited from native assistants about their own detailed knowledge of folk botany may be seen in Figure 2 where Alonso describes the appearance of the leaves of the orange tree. A translation of Alonso's text points out a highly complex Tzeltal botanical vocabulary for describing plant morphology. He writes, "The leaves of the true orange are green, glabrous, and small, perhaps half a "č'utub" in length [the measurement between the outstretched thumb and tip of the index finger], and about two thumb widths wide. There is a single large vein down the middle of each leaf. Leaves are lanceolate in shape, each showing a narrow, acuminate tip." This Mayan ethnophytological description, with the highly detailed use of descriptive adjectives for the major morphological areas of the leaf, is comparable to one that might have been prepared by a highly trained botanist.

The systematic nature of Alonso's characterizations of the use of plants in a number of culturally important ways is seen in his descriptions of the production of animal traps. Figures 3 and 4 are detailed drawings of two of the seven types of animal traps commonly produced in Alonso's area. The two traps shown, one for the capture of rabbits ("pehc'ul te'el t'ul") and the other for small birds ("pehc'ul te'el te'tikilmut"), are labeled to indicate the name and function of each part of the contraption. Each drawing is accompanied by texts of several pages describing the species of plants appropriate for the various parts of each trap and the several steps involved in the production of these ingenious devices.

Aguaruna and Huambisa ethnobotanical research

Let me now turn to the ethnobotanical research conducted among the Jivaroan-speaking Aguaruna and Huambisa in Amazonian Peru. Following the tradition begun in Chiapas, I trained two young assistants, Ernesto Ancuash and Rubio Kayap, in basic field collecting techniques. In three seasons (1972 to 1976), the two prepared over 4000 specimens to my 1200 for the same period. In 1979 to 1980, when ethnobotanical work was begun among the Huambisa, two additional collectors were brought into the project, Victor Huashikat and Santiago Tunqui. In a period of six months, the two had prepared more than 5000 collections in sets of five. More than three-fourths of the nearly 10,000 numbers that have resulted as part of the First and Second Ethnobiological Expeditions of the University of California to the Upper Marañón River have been made by Jivaroan collectors.

As with the Tzeltal Maya, the Aguaruna and Huambisa ethnobotanists kept their own collection notebooks. Figure 5 illustrates a page from the early collection

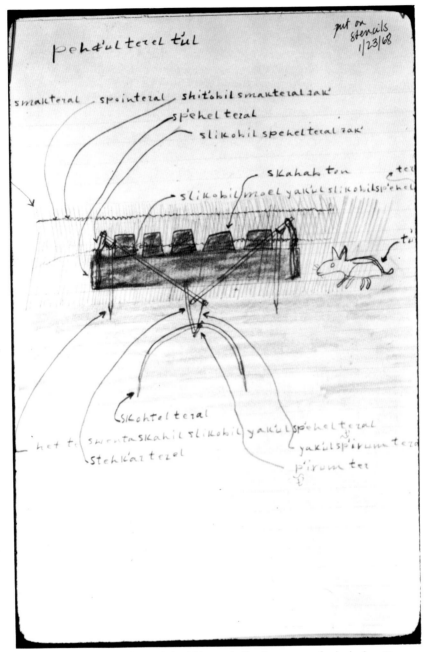

FIG. 3. Drawing of a Tzeltal rabbit trap indicating names of principal parts.

notes of Rubio Kayap. The entry provides both good botanical as well as eth-nobotanical information. Freely translated, we note "curaray [its Aguaruna name, a widespread word, borrowed from the Spanish, for *Potalia amara*], a tree of 2 meters in height, leaves fine and green in color, measuring 51 by 21 cm., flower white, small. Grows in monte (ajashbau or secondary forest), taken on the banks

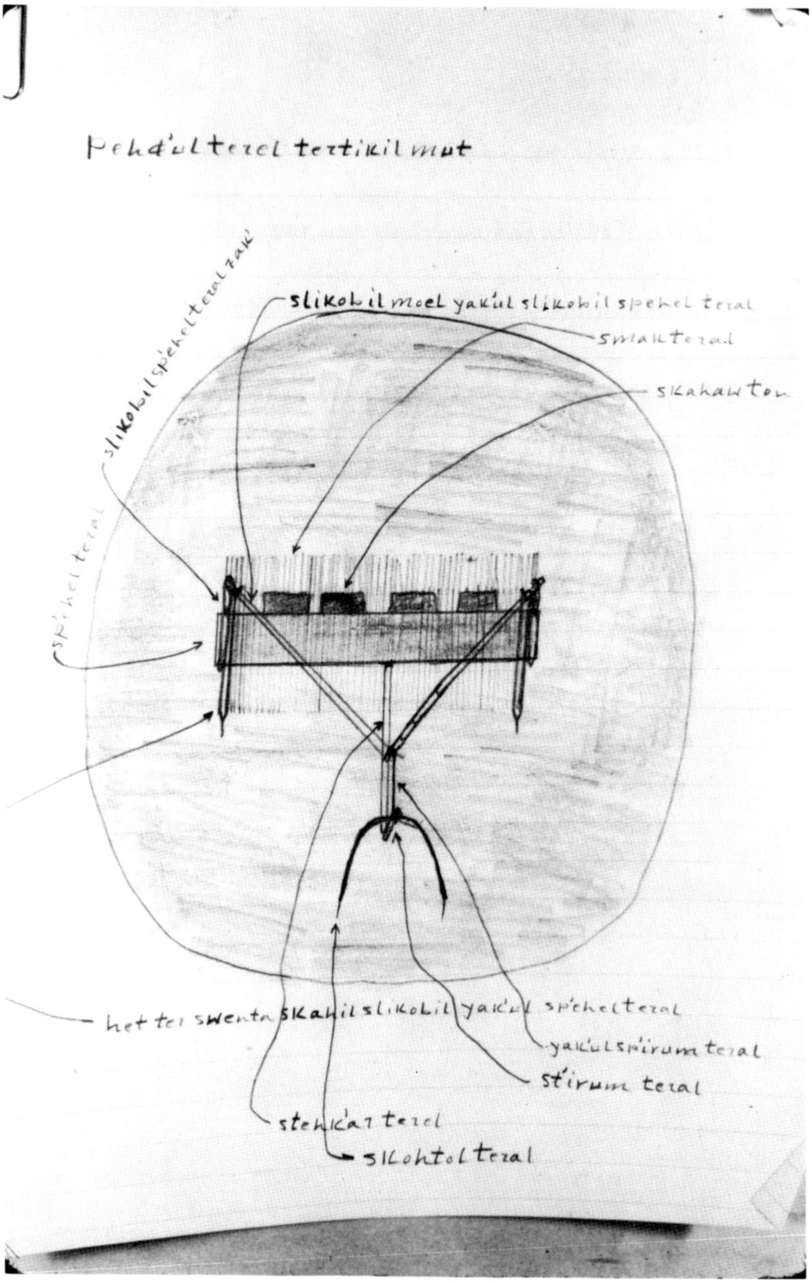

FIG. 4. Drawing of a Tzeltal bird trap indicating names of principal parts.

of Huampami Creek, 510 feet elevation, bark of the trunk used as a medicine [later determined to be important as a presumed snakebite remedy]."

Within a short while, a more efficient collection notebook was utilized in the research, following the suggestion of Theodore Dudley of the U.S. Arboretum. With this method, native collectors used field-label note pads supplied with carbon

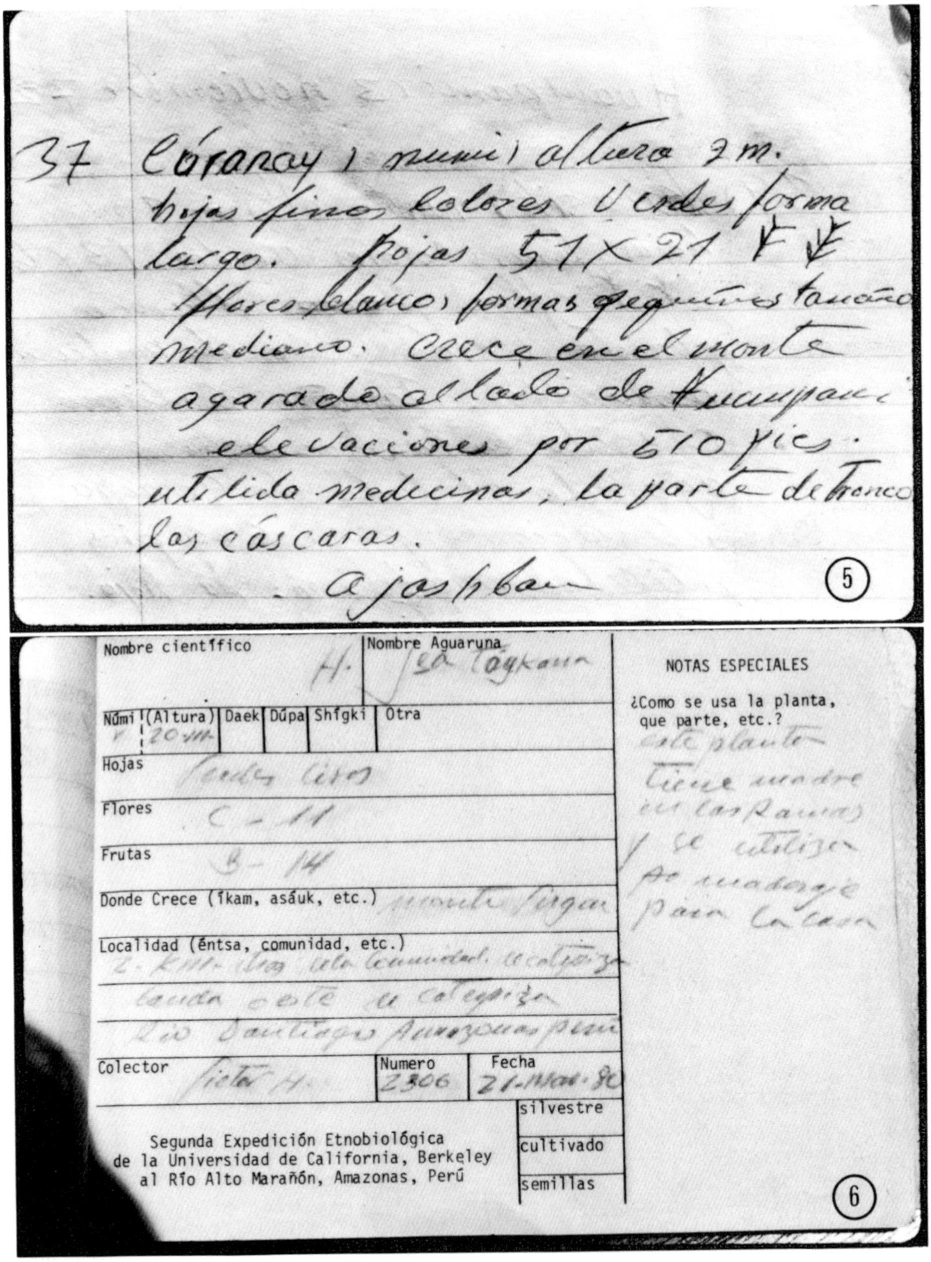

FIGS. 5–6. 5. A page from the botanical notebook of Rubio Kayap. 6. A page from the botanical notebook of Victor Huashikat.

paper and prepared their own field labels in duplicate. One label was kept with one sheet of the plant specimens of a particular collection number, the other kept as a carbon in the field-label notepad. This proved to be a major time-saver and is especially recommended for field situations where the native collectors are not yet comfortable with writing detailed descriptions on their own. Figure 6 shows a page from the collection notebook of Victor Huashikat. The label provides a place for the native name (in this case, "jea tagkana"). Label sets were printed for use with Aguaruna informants and the "H" indicates that the collector is a Huambisa. The codes indicated for "flores" and "frutas" (C-11 and B-14) are coordinates from a standard color chart (Berlin & Kay, 1969) that each collector used to determine color characters of plant specimens. Color notes were especially crucial because all collections were preserved with formaldehyde. In the section of the label marked "notas especiales," Victor writes that the tree harbors stinging ants ("tiene madre en las ramas," a common regional expression used to note this particular plant/animal association).

Aguaruna and Huambisa assistants were taught to make general botanical collections of any and all plants in flower and fruit that they might come upon on any collecting trip. Their collections have resulted in the discovery of a number of species new to Peru as well as new to science, some of particular economic significance. A good example is the tree called "chipa" in Aguaruna and Huambisa which, as a member of the resin-rich Burseraceae, provides the favored slip for Aguaruna and Huambisa pottery vessels. When specimens of chipa were examined by Duncan Porter, he deemed them sufficiently different from species of *Protium,* which it resembles, to describe it as a new genus. The specific epithet of the new taxon will commemorate the native name "chipa" which has been used by Jivaroan ethnobotanists for centuries and which, therefore, is a fitting tribute to their ethnobotanical expertise.

Botanists may ask how adequately can native assistants, functioning as independent botanical collectors, sample a particular local flora. Without specific training and direction, these assistants will probably overlook families of non-flowering plants. For example, I am certain that the ferns, especially, are poorly represented in our Rio Marañón collections. Much the same can be said of families of small, inconspicuous, herbaceous plants such as some orchids, peperomias, and even bromeliads.

Nonetheless, systematic efforts to collect all plant taxa that have been given names by a particular ethnic group will result in rather good coverage of the dominant species of a local flora. To my knowledge, this suggestion was first articulated over 50 years ago by one of the greatest ethnobotanists, Harley Harris Bartlett. In an infrequently cited monograph, *A Method of Procedure for Field Work in Tropical American Phytogeography,* Barlett wrote (1936: 8),

> The method involves the assumption that natives who have occupied an area for generations will have names for the dominant species in the flora, and that persistence in collecting the names and finding out how they are applied will result in accounting for most of the species which should be taken into consideration if plant associations are to be recognized in the phytogeographic reconnaissance of botanically new country.

Cyrus Lundell, a colleague of Bartlett, states (pers. comm.) in a similar vein that ". . . the only way to get complete coverage is to have a native resident to collect throughout the seasons."

This suggestion has been strictly followed in our Peruvian work where collaborators were instructed to produce comprehensive written inventories of all recognized, named plant taxa in the local flora and then to monitor carefully that list as collections were made, making efforts to procure adequate, determinable collections of every name on the list. These inventories of plant names thus serve as targets for future collecting efforts and provide some indication of just what remains to be done in a particular region.

While no published floras exist for the Upper Marañón River region of Peru with which our collections might be compared, it is interesting to compare gross numbers of species represented in our collections with the figures reported for two other well-botanized regions of the Neotropics. Our Peruvian collections total 1200 species in 125 families. This number is close to the 1300 species (of which 100+ were ferns) recorded for Barro Colorado Island (Croat, 1978). The flora of Río Palenque, Ecuador is comprised of 1112 species of higher plants (Dodson & Gentry, 1978). The similarity in total numbers of species represented in the three studies is striking and suggests that our native associates are approaching rather complete local coverage.

As part of basic botanical collecting, our Jivaroan assistants were asked to produce data on a broad gamut of ethnobotanical topics comparable to those focused on in the Tzeltal Mayan work (e.g., horticulture, gathering, house construction, ethnomedicine). These data reveal a rich inventory of plant-derived natural resources whose full potential is only partially understood by Western science. Our Jivaroan collaborators systematically cultivate 78 species, 48 of which are major food plants. As Boster (1984) indicates, numerous cultivars, such as manioc, have undergone deliberate horticultural manipulation leading to the development of scores of distinct varietal forms. Native assistants have provided major insights into this domestication process.

Another 120 species of wild plants are important for their edible fruits, and another 600 are important for their utility in house construction, cordage and fibers, poisons, tools, weapons, pottery production, dyes, and personal adornment. Ethnomedicinal aspects of plant use are seen in a number of areas such as preventive dental hygiene, as suggested in the paper by the Lewises (1984) and in fertility regulation (E. A. Berlin, in press). As one begins to appreciate the benefits of involving fully native assistants in these cultural aspects of ethnobotanical research, one is reminded once again of the validity of Carl Sauer's dictum on natural resources: "The environment can only be described in terms of the knowledge and preferences of the occupying persons; 'natural resources' are in fact cultural appraisals" (1952: 8).

Conclusions

I would like to close with an aside concerning the future collaborative endeavors of botanists and anthropologists in the field of ethnobotany which, I hope, will not be taken as presumptuous. Over the last 20 years, a number of changes have occurred in the mutual research program developed between field anthropologists interested in the ethnographic importance of a society's knowledge of plants and systematic botanists who have chosen to become concerned with at least some aspects of anthropologically focused issues in ethnobotany. This collaboration has led, on the parts of anthropologists, to the development of a recognition of the importance of making good botanical specimens, of collecting thorough label data, and even of making special efforts to collect intensively in some particularly favored genera that are known to be of special interest to a botanical colleague.

A number of botanists, on the other side, have incorporated anthropological themes into their own work, as exemplified most vividly by the contributions of botanists to the present volume. Nevertheless, many field botanists, who have had and will continue to have the opportunity to work in areas inhabited by native Americans, have not, unfortunately, taken advantage of these unique field situations to gather basic ethnobotanical data as part of the standard process of botanical collecting. Our experience has shown that minimal efforts devoted to training native collaborators can yield bountiful results. Not only can native assistants be taught to collect plant specimens, thus greatly increasing the botanical coverage for a particular region, but they can also be encouraged to produce the kinds of invaluable information on native knowledge of the plant world and of the application of this knowledge in daily life that I have briefly outlined in this paper. The benefits of widening the scope of routine botanizing in the American tropics to include these kinds of ethnobotanical data is underscored by the rapid and often deleterious social and ecological transformation of the neotropics that is currently sweeping the face of tropical America today. Although the situation may be changing, currently there are more trained tropical botanists than there are anthropologists with an interest in ethnobotany. Many of these botanists have seen and worked with more groups of American Indians than most practicing field ethnographers. If extensive ethnobotanical data are to be collected in the American tropics before the turn of the century, the work in large part must be coordinated by modern botanists who will take up the challenge of including native botanists as an integral part of their on-going research programs.

Literature Cited

Bartlett, H. H. 1936. A method of procedure for field work in tropical American phytogeography. Pages 1–25, Miscellaneous Paper No. 1. *In:* Botany of the Maya area. Carnegie Institute of Washington, Publication No. 461, Washington, DC.

Berlin, B. 1962. Esbozo de la fonología de Tzeltal de Tenejapa, Chiapas. Estudios de la Cultura Maya 2: 17–36.

———. 1968. Tzeltal numerical classifiers. A study in ethnographic semantics. Mouton and Company, The Hague, The Netherlands.

——— & **P. Kay.** 1969. Basic color terms: Their universality and evolution. University of California Press, Berkeley, California.

Berlin, E. A. In press. Fertility regulating mechanisms among the Aguaruna Jívaro. *In:* L. Newman, editor. Studies in indigenous fertility regulation. Rutgers University Press, New Brunswick, New Jersey.

Boster, J. S. 1984. Classification, cultivation, and selection of Aguaruna cultivars of *Manihot esculenta* (Euphorbiaceae). *In:* G. T. Prance & J. A. Kallunki, editors. Ethnobotany in the Neotropics. Adv. Econ. Bot. 1.

Croat, T. 1978. Flora of Barro Colorado Island. Stanford University Press, Stanford, California.

Dodson, C. H. & A. H. Gentry. 1978. Flora of the Río Palenque Science Center, Los Ríos, Ecuador. Selbyana, Volume 4.

Lewis, W. & M. P. F. Elvin-Lewis. 1984. Plants and dental care among the Jívaro of the Upper Amazon Basin. *In:* G. T. Prance & J. A. Kallunki, editors. Ethnobotany in the Neotropics. Adv. Econ. Bot. 1.

Sauer, C. 1952. Agricultural origins and dispersals. American Geographical Society, New York.

Steenis-Kruseman, M. J. van. 1974. Malaysian plant collectors and collections. Supplement 2. Page I–CXV. *In:* Flora Malesiana 8(1). Noordhoof International Publishing, Leyden, The Netherlands.

Wormersley, J. S. 1969. Plant collecting for anthropologists, geographers, and ecologists. Botany Bulletin No. 2, Department of Forests, Administration of Papua and New Guinea, Port Moresby, Papua New Guinea.

Classification, Cultivation, and Selection of Aguaruna Cultivars of Manihot esculenta (Euphorbiaceae)

James Shilts Boster

This paper has three principal objectives. The first is to describe how Aguaruna manioc cultivars are lost, found, and maintained. This description provides data on indigenous practices of manioc selection comparable to that provided by Conklin (1957) for rice (*Oryza sativa*), by Yen (1968) for sweet potatoes (*Ipomoea batatas*), and by Brush et al. (1981) for Andean potatoes (*Solanum tuberosum*). The second objective is to address the problem of the loss of genetic diversity of important crops. I stress, as Brush and his collaborators (1981) do, that an understanding of the social context of crop selection is a key to dealing with this serious problem. The third objective of this paper is to provide background information to support the claim that Aguaruna manioc cultivars have been selected for their perceptual distinctiveness. I will develop this idea more thoroughly elsewhere (Boster, unpubl.).

This paper recognizes three types of cultivar selection: perceptual, cultural, and natural. By perceptual selection, I mean the addition to an inventory of those cultivars that can be distinguished from other similar appearing cultivars. Such selection for perceptual distinctiveness is a preliminary step in the overall process of selecting a manioc cultivar. If a cultivar can be perceptually distinguished, its fate depends on its utility and hardiness. Cultural selection refers to the selection of cultivars for characters affecting the use of the plant while natural selection culls those plants that cannot survive the hazards of the environment. This paper will primarily consider these last two modes of selection: deliberate cultural selection for desirable properties related to the consumption of the cultivars and natural selection for resistance to pathogens, pests, and other environmental hazards. The selection of manioc cultivars is only one aspect of the dynamics of the

Aguaruna manioc cultivar inventory; this paper also describes the loss and introduction of cultivars in this system. Finally the paper describes the ways in which
this dynamic system is reflected in Aguaruna manioc nomenclature.

The Aguaruna Jívaro live in humid tropical forest on the hilly rim of the Amazon
basin in Peru. They subsist by gardening, fishing, hunting, and collecting forest
products. Women are the principal gardeners (Fig. 1) while men are the principal
hunters. Although many crops are grown by the Aguaruna, manioc is by far the
most important. It contributes more than half of the calories in the diet (Berlin
& Berlin, 1978: 26) and makes up more than three quarters of the individual
plants in the gardens. In fact, the diversity of distinct cultivars of manioc maintained in the polycropped gardens is greater than the diversity of distinct crops
(Boster, 1983). The starchy root is prepared for consumption by boiling or roasting
or by fermenting for beer.

The field research was conducted between July 1977 and September 1978 as
part of the Segunda Expedición Etnobiológica al Río Alto Marañón, led by Brent
Berlin. My methods of data collection included asking informants to identify by
native name manioc plants growing in experimental gardens, conducting stan-
dardized interviews to determine the culturally important properties of the manioc
cultivars and the flow of cultivars through the community, taking transects of
native gardens to determine the frequencies of the various manioc cultivars and
other cultivated plants, recording of yields of garden production, engaging in in-
depth, open-ended interviews on cultivation practices, and simply observing. The
data were collected by myself and by trained, literate Aguaruna and Huambisa
research assistants.

Selection and maintenance of cultivars

Aguaruna horticultural practices are typical of a lowland South American swidden system (cf. Carneiro, 1964; Denevan, 1971; Kensinger, 1975; Leeds, 1961).
Well drained, sandy, stream-side sites covered with mature secondary growth are
preferred for gardens because they can support the largest inventory of crops with
the least effort. Gardens are cleared with the use of axes and machetes. Often the
felled trees and slash are burned, but sometimes the slash is merely dumped at
the edge of the garden. The garden is prepared for planting by loosening the earth
with a palm-wood digging stick and inserting the manioc planting stems into the
resulting mounds. Most of the planting stems for a new garden are taken from
the cultivator's other gardens. Many of these are acquired in the daily harvesting
of roots; a single healthy plant when harvested can supply from five to twenty
planting stems. Additional planting material is harvested from plants that have
multiple trunks sprouting from the same original planted stem. One of these can
be used for planting material without harm to the parent plant. At seven to ten
months, the manioc reaches maturity and the cultivator begins a daily routine of
weeding small sections of the garden, harvesting the plants in the weeded areas,
and replanting almost the entire length of each stem in the mounds created by
harvesting.

Planting of a new garden is the occasion of the most intense selection of cultivars.
Some effort is made to keep a good mix of cultivars in the garden. If a cultivar
is judged to be under-represented in the older garden, as many planting stems as
possible are cut from the plants that do exist to reestablish a balance. If a desired
cultivar cannot be found in the older garden, it is often requested from a kins-
woman, usually a mother or sister. The judgment of whether a cultivar is under-
represented depends on the ability of the cultivator to identify the cultivar in

FIG. 1. An Aguaruna woman peeling a manioc root while seated in her garden in Amazonas, Peru. (Photo by Brent Berlin.)

question. Women are making planting decisions on the basis of the properties of the taxonomic *categories* which they recognize; without a classification scheme it would be extremely difficult to make systematic associations between the physical specimens and the culturally valued properties. Selection without a classification scheme would require that the cultivator observe and remember the performance of each of thousands of plants in the garden. Though there are instances in which cultivators take note of exceptional individual plants, for the most part the cultivator's evaluations of the performance and properties of cultivars, decisions of which cultivars to grow, and discussion and exchange of cultivars with others are done by reference to named taxonomic categories (cultivars, in this case) instead of individuals (cf. Brush et al., 1981).

To investigate the criteria used by Jivaroan women in the deliberate selection of manioc cultivars, a standard interview was administered to a total of 217 Jivaroan women from nine native communities. The interview posed thirty questions covering growth rate and yield, resistance to flood and wind damage, culinary properties, and problems in cultivation, food preparation, and storage. These questions could all be answered by either a list of manioc cultivars or claims of ignorance of the answer. The interview questions are presented in Appendix I.

All interview responses of one local group ($N = 70$) were further analyzed (Boster, unpubl.). The objectives of this analysis were to determine 1) what the culturally important properties attributed to the manioc cultivars are, 2) how these properties are associated with one another, and 3) which properties are most important in the decision to plant more of some cultivars than of others.

The pursuit of the first objective resulted in Table I, which shows the number of times each of the manioc cultivars was named by this group of 70 women in response to each of the interview questions.

The second objective was met by examining the pattern of correlations between the properties attributed to the cultivars. Many of the culturally important properties of the manioc cultivars fall into two clusters. These two clusters reflect the contrast between the properties typical of those cultivars used as table vegetables and those used for beer making. In other words, when choosing desirable properties of cultivars, Aguaruna women typically cannot drink their manioc and eat it too. Root color is the most important determinant of the use of the cultivar: Aguaruna women generally make beer from the yellow-fleshed cultivars and boil or roast the white-fleshed cultivars. A summary of the women's responses to the interview questions indicates that the yellow-fleshed beer-making cultivars, in comparison to the white-fleshed eating cultivars, tend to grow more rapidly, produce larger roots, have more fiber, have a greater tendency to get hard, and rot more rapidly after harvest. It should be emphasized that these generalizations about the differences between the beer-making and eating cultivars are merely tendencies, not absolutes. For example, although the beer-making cultivars tend to have higher yields than the eating cultivars, the two cultivars most commonly planted for eating yield more than most beer-making cultivars. In addition, some cultivars prized as table vegetables can be used for making beer and vice versa.

To infer how Aguaruna decide the amount of each cultivar to plant, the third objective, I compared the properties attributed to the different cultivars with the relative frequencies of the cultivars. The relative frequencies of the manioc cultivars were determined by counting the numbers of manioc cultivars and all other crops in a 4 m by 30 m transect through the interview informants' gardens (Boster, 1983: 49). An average of 160 manioc plants representing an average of 12 locally recognized cultivars were identified by the owner of each garden sampled. There were probably many cultivars in the gardens that were not encountered in the sample area, which only covered about 5% of the typical garden.

A multiple regression was then performed using the frequency of occurrence of each cultivar in each woman's garden as the dependent variable and that woman's mention of the cultivar in answer to the interview questions as a series of binary independent variables. The results of this analysis can be summarized as follows:

Table I

Summary of interview responses and garden transect data[a]

Name	A	B	C	Question number																												
				1	2	3	4	5	6	7	8	9	10	11	12	13	14	15	16	17	18	19	20	21	22	23	24	25	26	27	28	29	30
paúm	0.015	3460	68	64	0	52	15	12	2	56	45	18	48	21	49	2	38	8	1	13	10	27	15	49	4	43	3	27	15	27	18	8	15
ujákag	0.016	2417	71	60	1	48	15	9	2	43	30	16	35	16	32	7	24	20	6	15	5	25	15	35	7	28	9	31	10	26	11	12	16
ipák	0.969	1385	65	66	1	40	19	8	4	52	57	11	22	12	4	60	30	7	8	13	12	6	36	4	49	20	26	23	20	40	4	25	24
yakia	0.032	919	59	62	1	36	7	7	1	37	23	21	22	8	21	8	26	3	0	8	3	37	11	33	8	27	6	20	7	14	10	9	15
sujíknum	0.984	378	43	65	1	25	49	30	45	2	0	0	52	18	28	3	14	9	2	1	3	9	21	18	7	24	7	18	10	12	23	13	17
shaámpig	0.053	291	35	56	3	20	2	5	1	18	17	9	14	5	17	8	12	5	7	4	5	7	23	5	18	13	15	3	4	8	5	5	8
puyám	0.949	268	43	59	1	32	36	23	32	3	1	1	25	11	24	5	14	8	3	4	1	11	8	19	2	10	4	12	9	3	12	3	5
kanúsag	0.101	169	35	59	2	19	8	16	3	17	12	3	17	24	6	11	6	6	9	6	13	4	15	3	7	8	6	6	4	7	5	4	6
shímpu	0.024	160	21	41	0	10	4	5	5	12	6	3	9	12	4	3	10	6	5	3	4	1	9	4	10	2	5	4	4	4	3	5	6
takítag	0.829	159	24	41	0	14	10	12	15	3	2	0	11	7	10	2	1	14	1	3	5	3	6	9	2	5	4	5	3	0	3	0	6
paágmash	0.940	140	19	50	1	13	12	12	9	5	2	0	7	5	14	8	9	4	9	6	4	21	9	8	8	6	6	9	7	6	1	10	3
kakugpátin	0.928	130	20	28	1	12	5	4	2	5	0	6	2	1	8	1	1	2	1	2	0	1	3	2	6	2	4	2	4	4	4	4	2
chikím	0.023	110	15	43	0	13	4	3	4	10	6	5	8	10	20	5	5	4	4	8	4	8	15	1	13	1	1	8	5	6	9	1	3
ikítus	0.923	100	11	13	0	3	4	6	1	4	3	0	1	4	1	2	3	0	0	2	3	1	5	2	1	2	1	0	1	0	0	0	1
antúk	1.000	92	16	26	1	7	5	0	1	12	11	1	7	2	0	11	4	2	1	6	1	1	6	2	11	2	3	4	4	8	0	1	3
jijuántam	0.047	88	20	42	2	12	6	3	1	8	19	4	3	16	4	10	8	10	5	2	6	2	7	4	7	4	5	7	4	5	8	4	2
wagkám	0.937	73	15	32	1	11	8	9	12	3	3	2	19	3	2	9	13	2	5	1	3	0	9	2	15	5	2	7	9	9	3	10	2
nantujá yutú	0.166	58	5	6	0	3	1	0	0	1	0	3	1	1	0	1	0	1	0	0	0	3	0	1	0	1	0	0	1	0	0	0	1
ukaeyín	0.000	55	15	53	3	16	3	6	0	0	27	5	1	4	1	34	18	2	14	8	2	28	5	0	25	11	10	12	11	4	2	2	8
butuúm	0.840	55	12	25	2	8	2	4	4	4	0	2	4	4	3	5	2	0	2	0	0	3	0	3	4	3	5	2	1	2	1	1	2
dapím	0.619	50	8	21	2	4	4	4	2	0	2	1	2	1	2	2	0	5	4	2	0	0	2	2	1	1	0	3	0	1	2	2	1
klára	0.000	48	9	13	0	7	0	0	6	6	9	4	6	2	8	6	2	1	0	4	0	2	4	1	0	8	3	2	1	2	0	2	1
sujíktak	0.954	47	10	44	1	7	16	13	22	0	0	0	10	15	2	2	1	10	8	1	3	2	8	5	2	7	3	3	8	3	5	4	8
shimpím	0.052	46	11	38	4	11	5	1	0	12	0	0	2	5	5	10	12	0	5	8	9	3	9	2	16	9	2	4	8	5	4	1	5
báshu dáwe	0.000	45	11	22	0	2	0	0	1	0	1	0	0	1	1	1	1	3	1	2	0	1	1	0	3	1	0	0	1	0	1	0	0
suíg	0.758	44	9	29	1	5	8	8	7	3	0	7	3	3	3	3	0	4	0	2	3	0	2	3	2	0	0	3	0	2	0	0	2
taáma	0.500	44	2	16	1	1	1	1	1	0	1	0	1	9	0	0	1	2	1	0	1	0	3	0	2	0	1	1	0	0	1	0	1
batiátsag	0.111	43	8	18	0	1	1	3	2	1	2	0	0	0	0	5	2	0	1	1	1	3	1	4	4	1	2	1	2	0	0	0	1

Table I
Continued

Name	A	B	C	1	2	3	4	5	6	7	8	9	10	11	12	13	14	15	16	17	18	19	20	21	22	23	24	25	26	27	28	29	30
piampia	0.312	43	6	16	0	1	0	2	0	2	1	1	0	0	0	2	2	0	0	2	0	2	2	2	2	2	1	0	1	0	1	0	1
kigkísag	0.807	42	10	26	1	5	3	2	3	5	2	0	3	4	0	4	4	4	1	2	2	1	2	1	3	3	2	0	0	1	0	0	2
mún máma	0.120	41	8	25	0	9	2	1	1	5	1	0	7	2	2	2	4	2	1	0	0	1	1	2	2	0	0	0	2	0	2	0	0
chunúk	0.312	39	6	16	0	3	3	0	2	3	0	0	0	1	0	1	0	3	1	0	2	1	2	1	1	0	2	0	1	0	2	1	0
yusanía	0.086	37	11	23	0	3	3	2	2	3	0	0	0	3	0	1	1	1	0	1	3	1	3	2	3	2	2	3	1	1	3	0	1
atsásua	0.950	37	9	20	1	6	9	3	5	1	2	0	7	12	7	1	2	3	2	1	1	3	0	3	3	0	0	3	1	0	1	1	5
dabuúg	1.000	37	7	20	0	7	11	9	15	0	0	0	5	4	0	2	2	1	1	2	1	0	3	1	1	1	0	5	0	1	1	1	2
apág	0.722	37	8	18	0	7	1	1	3	1	1	0	4	3	2	2	1	3	1	0	1	0	2	2	2	3	2	0	3	2	1	0	1
sakém	0.892	31	5	28	2	5	8	4	7	2	1	0	3	3	0	2	2	2	2	4	3	2	3	3	3	2	5	1	5	0	2	1	2
wampúshig	0.066	29	7	30	1	2	0	2	0	3	1	1	2	5	0	1	2	2	2	0	0	1	1	2	2	2	1	1	2	0	0	0	1
patáku	0.653	28	11	26	2	6	6	7	6	2	2	0	3	3	2	4	2	2	1	1	3	0	2	2	2	4	0	0	0	2	0	4	1
tanaím	0.113	26	7	44	3	13	4	3	3	10	3	1	1	2	23	2	7	11	3	2	3	18	3	1	1	6	1	3	5	4	10	2	3
shuskugkía	0.500	25	5	20	0	1	3	3	3	2	2	0	0	8	0	3	0	2	3	0	0	0	4	0	2	2	2	2	3	0	2	1	1
tsápak	0.416	25	4	12	0	1	0	1	2	2	0	0	2	2	0	2	2	0	0	0	0	2	1	3	3	3	1	2	0	1	1	0	1
ushuúmchatai	0.074	24	6	27	1	4	0	3	0	2	0	0	0	3	0	2	2	2	2	2	2	0	4	0	5	0	1	4	2	1	2	0	1
shuín	0.300	23	7	10	0	4	2	0	2	2	0	0	2	1	1	1	1	2	1	2	0	1	3	3	3	2	0	0	0	0	1	2	0
yampítsag	0.850	21	6	20	1	9	2	1	1	2	0	0	1	1	0	4	4	1	2	1	1	3	3	2	2	0	0	2	2	0	0	0	1
tutúpik	0.100	20	4	10	0	3	0	2	0	3	1	0	2	0	8	1	1	0	0	0	0	2	3	0	2	3	0	0	0	0	0	3	1
kugkuín	0.944	19	6	18	1	4	1	0	0	5	5	0	5	1	0	3	6	0	1	0	1	1	1	1	1	2	0	0	0	0	1	0	0
untsumá	0.500	18	3	2	0	0	0	0	1	1	0	0	0	0	0	0	0	0	2	0	0	0	0	0	1	0	0	0	0	0	0	0	0
ipág	0.090	16	6	22	1	5	0	1	0	0	0	0	0	1	10	4	1	2	2	0	3	5	3	7	4	0	1	2	1	1	1	0	0
nampuín	0.583	16	6	12	2	2	5	4	2	1	0	0	3	2	1	1	2	0	0	0	0	0	0	2	1	0	2	0	0	1	0	0	2

a The numbers in column A indicate the proportion of times that the variety was described by interview informants as yellow, rather than white. Thus the higher values are associated with the yellow varieties while the lower values are associated with the white-fleshed varieties used for boiling and roasting. The figures in column B are the total numbers of individual plants of each variety found in all 74 gardens; 11,857 manioc plants were inventoried in all. The figures in column C are the numbers of gardens out of the total 74 in which each variety was found. The figures in the following 30 columns are the number of times that the variety was mentioned in response to the 30 interview questions (Table I).

Aguaruna women grow some of all the cultivars available but will plant more of a cultivar if the cultivar can be harvested rapidly, if it is good when boiled and eaten, if it does not rot rapidly after harvest, and if it produces large roots.

One unexpected finding of this research project was that environmental response characteristics of manioc seemed to be of little concern to the Aguaruna. Most of my informants regarded my questions about cultivar differences in flood, wind, and pest resistance as silly, even though these are regarded as serious threats to the survival of a garden. The most common answer to these questions was that all cultivars would be destroyed in a severe flood, a wind storm, or an attack by leaf cutter ants. Similarly, I was disappointed in my expectation that the Aguaruna would match the physiological characteristics of the cultivars to specific soil conditions as reported by Kensinger (1975) for Cashinahua manioc cultivation. Although some women planted the different cultivars in segregated clumps, it appears that the motive for doing so was to help distinguish the cultivars rather than to take advantage of garden microenvironments.

Nevertheless, the cultivars apparently did differ in their response to soil conditions. As part of this study, I asked cultivators to keep track of the yields of all plants harvested from their gardens for a period of six months. One participant in this project discovered that some cultivars grow well in the sandy alluvial soils on river islands and do poorly on other sites while other cultivars yield about the same regardless of soil conditions. I conclude from this discovery that there are significant differences between the cultivars in their responses to environmental factors and that the Aguaruna are generally unaware of or unconcerned with these differences. Two possible explanations of why the Aguaruna do not worry about the cultivars' responses to the environment are 1) that this is an aspect of selection that is efficiently taken care of by natural processes or 2) that deliberate selection for yield indirectly selects for resistance to a wide spectrum of environmental conditions.

Loss of cultivars

Interviews with informants indicated there is a steady loss of cultivars from the total inventory. This loss can take place through rejection of individual cultivars, accidental loss, and large scale abandonment. Cultivars are rejected outright only under extremely unusual circumstances. Less desirable cultivars are likely to be planted less frequently rather than abandoned, so that the cultivator retains the option to decide later that she wants it. The rarer cultivars are more vulnerable to accidental loss through random events, such as choking by weeds or attack by pests.

It appears that the greatest loss of manioc cultivars is through wholesale abandonment, either directly through loss of the planting material itself or indirectly through loss of the knowledge of the differences between the cultivars. Cultural contact with the national society seems to have had both effects. In about 1950, Catholic and Protestant missionaries began introducing schools to the Aguaruna and the neighboring Huambisa Jívaro. In forming new communities around the schools, the missionaries drew the inhabitants of the widely scattered households down from the headwaters of the smaller creeks and concentrated them on the banks of the major rivers. My Aguaruna informants complained that many cultivars were lost in the move, abandoned in the upstream gardens, because the planting material was too cumbersome to transport. The severity of the loss of cultivars seems to be a function of the difficulty and precipitousness of the relocation. The Huambisa of the Río Santiago have much smaller inventories of manioc cultivars than the Aguaruna probably because, according to Huambisa oral histories, the distance that they had to travel to the new communities was much longer and over rougher terrain and because their movement was made more or less simultaneously by many people to destinations with little or no previous native occupation.

The introduction of schools also seems to have promoted substantial loss of knowledge about the distinct cultivars: by attending school, girls were deprived of their education with their mothers and sisters in the manioc gardens (Boster, in press). Because the maintenance of a diversity of manioc cultivars depends on the ability of the cultivator to distinguish them, in the long term the loss of knowledge of manioc cultivars will result in the loss of the genetic diversity of manioc.

Introduction of cultivars

This loss of cultivars is partly compensated by the influx of new cultivars through exchange, some over a great distance. Certain cultivars were said to come from as far as Iquitos and Pucallpa, many hundreds of miles away, by trade with relatives downstream on the Río Marañón. However, most of this exchange is within the community. Older women are especially proud of their knowledge and ownership of a range of different manioc cultivars. The gardens of these women served as a source of rare cultivars for other women in the community.

There is evidence that the exchange of manioc cultivars is extensive. As a part of the interview, I asked informants to state who had given them each of their cultivars. This data allowed me to discover the proportion of the women who directly or indirectly exchanged cultivars. Although the proportion of immediate exchanges between the 70 informants from five communities was only 11%, 94% of the informants were potential exchangers of planting material through three intermediaries or less. This means that any promising cultivars could rapidly and efficiently be diffused through the social network.

A second source of new cultivars is through the nurturing of volunteer manioc seedlings. According to my informants, sometimes manioc fruits in abandoned gardens fall onto the ground and lie dormant through the period of secondary forest succession. When the site is cleared again for a garden, the seeds germinate and grow. These young plants, discovered after the clearing is finished and before planting has begun, are recognized as volunteer seedlings. My informants emphasized that these volunteers originated from seeds instead of from stems because stems would rot much too soon to survive succession. They also reported that such volunteer plants which had been dug up were found to have originated from seed. My informants found it difficult to estimate the frequency of such a rare event, but their guesses averaged roughly one volunteer seedling for every two gardens cleared. Though this frequency may seem extremely low, it is the most important source of new genetic combinations because no informants reported deliberate planting of manioc seed. Some women on finding a volunteer seedling note its location and wait until the plant is mature. The roots are then cooked and tested for their bitterness. If they are bitter, the stem is discarded. If the roots are not bitter, then the stems are replanted and the clone treated like any other cultivar. These cultivars are called "yagkují" (flower) or "tsapaínu" (sprout) to indicate their origins as volunteer seedlings.

It appears that selection for perceptual distinctiveness operates most intensely when a cultivar is first introduced. The introduced cultivar is likely to be regarded as an example of one of the locally occurring cultivars if it too closely resembles one. For example, I planted in an experimental garden a stem I was told was "waségmish" (Fig. 2c), a cultivar that came from a neighboring drainage. Most informants identified it as "mamáyakem" (Fig. 2d), a local cultivar it resembled. Both cultivars have large dark obovate leaves, green petioles, and silver stems. Only the woman who had given me the stem in the first place 'correctly' identified it. It did not appear that she was using "mamáyakem" and "waségmish" synonymously because she agreed with other informants in her identification of "mamáyakem." Although clearly one can learn to distinguish the cultivars, as the women who gave it to me did, I believe the limited diffusion of "waségmish" is a partial result of its similarity to the locally occurring "mamáyakem." A contrasting case is that of "yarína máma." This cultivar had been recently introduced and was cultivated and correctly identified by only one small group of related women. Since "yarína máma," with its purplish red petioles and striking dark young foliage, does not strongly resemble any local cultivar, other informants were much more likely to recognize that they had never seen it before. Here, as in the rest of human experience, recognition of one's own ignorance is a prerequisite to learning.

This example illustrates that the perception of folk biological categories is by no means fixed: highly skilled and motivated informants with extensive experience with the organisms can make exceedingly subtle discriminations between groups of organisms. The fact that the owner of "waségmish" was able to distinguish extremely similar cultivars which other women regarded as identical indicates that the community of cultivators is inherently heterogeneous in motivation, experience, and skill. It appears that the maintenance of a heterogeneous inventory of cultivars is partly a result of the heterogeneity of the cultivators. This parallels Harlan's observation (1975; cited in Brush et al., 1981) that human cultural diversity promotes crop genetic diversity. The process of selection for perceptual distinctiveness is determined not so much by how carefully the local experts can attend to the cultivars but by the amount of effort the range of typical cultivators are willing to expend. Cultivars that demand careful attention to be discriminated will not diffuse as rapidly, all things being equal, as those which are more readily distinguished. Lack of perceptual distinctiveness does not prevent the diffusion of a cultivar but merely lowers the probability of its diffusion.

The outcome of this process of introduction, selective maintenance, and random loss is that there is a core of common, widely shared, and widely known cultivars and a much larger number of rarer cultivars known only by small numbers of women. This is evident in the great disparities in the frequencies of mention of the different cultivars. None of the manioc cultivars were mentioned by all the Aguaruna women interviewed, eighteen cultivars were mentioned by more than half the women, 127 were mentioned by three or more women, and more than 700 distinct cultivar names (many of them synonyms) were mentioned in total. The maintenance of different inventories of cultivars by different women increases the total diversity sustained by the community as a whole.

Aguaruna manioc nomenclature

The individual histories of the manioc cultivars are reflected in their names. There are at least four major sources of Aguaruna manioc cultivar names: folk genera of plants and animals, descriptions of characteristics of the cultivar, place of presumed origin, and name of the person who introduced it. The latter two classes of names are evidence of a productive naming system, a way of incorporating new cultivars into the classification scheme. A cultivar is often named

FIG. 2. Leaves from specimens growing in the first of two experimental gardens. Their Aguaruna names are a) "puyám," b) "magkám," c) "waségmish," and d) "mamáyakem."

after the person who introduced it to the community, particularly if the introduction is recent. This practice might be regarded as analogous to the granting of a patent to an inventor; it is a means of recognizing individual initiative. Once these cultivars gain broader acceptance, they are often named after their place of origin.

The cultivars named after genera of plants and animals or some characteristic of the plant represent the core of the most important and most widely shared cultivars. When I asked my informants where these cultivars had come from and who had given them, a frequent response was that the cultivars had always been here and that "núgkui," the mythical donor of all cultivated plants, had given them. Cultivars named after other folk genera of plants sometimes shared some characteristic with it. "Chikím" (*Maranta* arrowroot manioc) and "sakém" (*Euterpe* palm manioc) both had narrow leaflets that resembled those of the plants they were named after. "Dapím" (snake manioc) also had long narrow leaves. "Báshu dáwe" (guan foot manioc) was so named because of the pattern of branching of roots from the stem. However, it was often difficult to detect the similarity between the manioc cultivar and the source of the name, particularly when the source was an animal, for example "páki máma" (peccary manioc), "piampía máma" (sandpiper manioc), "yampítsag" (dove manioc), and "basuíg máma" (stink bug manioc).

The descriptive names were often references to the culinary properties of the cultivars. "Pushuúm" (foam manioc) and "pushuútin máma" (foam having manioc) were both named for the foamy head they gave to manioc beer, "ushuúmchatai" (meat hunger eliminator) was said to satisfy a craving for meat, and "kugkuín" (fragrant manioc) was named for its cooking aroma. Other names were morphological descriptions: "puyám" (thin manioc) (Fig. 2a), "tsegkém" (narrow manioc), and "wagkám" (wide manioc) (Fig. 2b) all referred to the shapes of the leaflets; "nampuín" (swollen manioc) referred to the shape of the root; "patáku máma" (wound around manioc) referred to the pattern in which the roots were wrapped around each other; and that of "uúmig" (blowgun manioc) referred to its straight, tall, unbranched trunk. Still other names were whimsical, bearing no obvious relation to characteristics of the plants. Such names include "sujíknum" (stingy stick) and "tunaím" (waterfall manioc). The problems of deciphering the meanings of manioc names were compounded by the great variation in the pronunciation of the names. For example, "ujákag" ("pull up" manioc?) could also be called "uják," "ujám," and "jáku tségkamu." Sometimes this variation in pronunciation involved semantic variation as well. For example, the cultivar "ushuúmchatai" (meat hunger eliminator manioc) was also called "ushúuwakin," "ushuúm," and "úshu máma." The meaning of the last variant pronunciation can be translated as *Caladium bicolor* manioc. It appears that the original meaning of this cultivar name became obscure and the variant pronunciations reflected different speakers' rederivations (or folk etymologies) of the term from different sources.

Conclusions

This paper has attempted to convey the idea that the maintenance of crop genetic diversity depends on the cultural practices of the cultivators. The description of these cultural practices has pictured the maintenance of an inventory of cultivars as a stochastic process influenced both by deliberate human choice and random chance. Aguaruna manioc nomenclature reflects and reinforces this complex process of cultivar selection and maintenance. Aguaruna selection of manioc cultivars involves both unconscious selection for combinations of characters that allow the cultivars to be perceptually distinguished and deliberate selection for desirable characters that affect the use of the plants. Yield is an important but not over-riding concern; other considerations include suitability of a cultivar as a table vegetable or for beer making, storage qualities, and rate of growth. Resistance to environmental hazards seems unimportant to the Aguaruna as a reason to choose to plant one cultivar more than another.

Cultivars are more commonly introduced through exchange than through the encouragement of volunteer seedlings. It is extremely unusual for a cultivar to be deliberately banned, but sometimes cultivars are lost by accident. The rarest cultivars are most vulnerable to accidental loss. Wholesale abandonment of the cultivar inventory is the most serious threat to this genetic resource. Instances of wholesale abandonment have usually been the result of contact with the national society, especially the introduction of schools. At least some of the missionaries who introduced schools to the Aguaruna are painfully aware of the ethical dilemma this situation presents. The Aguaruna need western education to adapt to the changing social environment caused by cultural contact. However, one of the costs of this formal education is the loss of opportunities to acquire traditional knowledge necessary to their adaptation to the natural environment. This illustrates one of the many dilemmas inherent in attempts to retain crop genetic diversity in the face of the erosion of traditional cultural practices.

Acknowledgments

This article is a modified version of a paper prepared for a symposium entitled "Ethnobotany in the Neotropics" held at the June 1983 Annual Meeting of the Society for Economic Botany at Miami University, Oxford, Ohio. The research was funded by National Science Foundation Grant number BNS 7916746, Brent Berlin principal investigator. Travel was supported by grants from the Tinker

Foundation administered through the Center for Latin American Studies at Berkeley. Additional support for the analysis of the data was provided by a University of Kentucky Summer Faculty Fellowship. I am grateful to Herbert Baker, Stephen Beckerman, Brent Berlin, Bill DeWalt, Jack Harlan, Jacquelyn Kallunki, Maria Lebrón, Ghillean Prance, Sara Quandt, and an anonymous reviewer for comments and criticism. I give special thanks to the editors of this volume: to Ghillean Prance for organizing the symposium and inviting me to participate, and to Jacquelyn Kallunki for gently and tenaciously prodding me to correct ambiguities and errors in the original manuscript. She often was able to state clearly and explicitly points I had been struggling to make. I would like to thank my Aguaruna and Huambisa research assistants and informants for their help in documenting their knowledge of manioc. I am also grateful to Herbert Baker and Jeff White for first introducing me to these issues and to Susan Stewart for helping in the search for relevant literature. Finally, I would like to thank my friend and mentor, Brent Berlin, for his collaboration in making this research possible. Any errors of fact or interpretation in the text are my own.

Literature Cited

Berlin, B. & E. A. Berlin. 1978. Etnobiología, subsistencia, y nutrición en una sociedád de la selva tropical: Los Aguaruna (Jíbaro). Pages 13–47. *In:* A. Chirif, editor. Salúd y nutrición en sociedades nativas. Centro de Investigación y Promoción Amazónica, Lima, Peru.

Boster, J. S. 1983. A comparison of the diversity of Jivaroan gardens with that of the tropical forest. Human Ecol. 11: 47–68.

———. In press. 'Requiem for the omniscient informant': There's life in the old girl yet. *In:* J. Dougherty, editor. Directions in cognitive anthropology. University of Illinois Press, Champaign, Illinois.

Brush, S. B., H. J. Carney & Z. Huamán. 1981. Dynamics of Andean potato agriculture. Econ. Bot. 35: 70–88.

Carneiro, R. 1964. Shifting cultivation among the Amahuaca of eastern Peru. (Beitrage Völkerkunde Sudameriküs.) Völkerkundliche Abhandlungen Band 1: 9–18.

Conklin, H. C. 1957. Hanunoo agriculture: A report on an integral system of shifting cultivation in the Philippines. Forest. Developm. Paper, No. 12. Food and Agriculture Organization of the United Nations, Rome, Italy.

Denevan, W. M. 1971. Campa subsistence in the Gran Pajonal, eastern Peru. Geogr. Rev. 61: 496–518.

Harlan, J. 1975. Our vanishing genetic resources. Science 188: 618–621.

Kensinger, K. M. 1975. Studying the Cashinahua. Pages 9–85. *In:* J. Dwyer, editor. The Cashinahua of eastern Peru. Studies in anthropology and material culture, Vol. 1. The Haffenreffer Museum of Anthropology, Brown University, Providence, Rhode Island.

Leeds, A. 1961. Yaruro incipient tropical forest horticulture: Possibilities and limits. Pages 13–46. *In:* J. Wilbert, editor. The evolution of horticultural systems in native South America, causes and consequences: A symposium. Antropológica, Suppl. Publ. No. 2, Caracas, Venezuela.

Yen, D. E. 1968. Natural and human selection in the Pacific sweet potato. Pages 387–412. *In:* E. T. Drake, editor. Evolution and environment. Yale University Press, New Haven, Connecticut.

Appendix I: Questions used in manioc interview

1 Amína ajagmínish wajſ yujúmka awa
1e What kinds of manioc are in your garden?
2 Tíkish núwa ajajínish, tíkish yujúmkash áwak

2e In the gardens of other women are there other kinds of manioc?

3 Wají yujúmka ajagmínish imá kuáshtash ukuaúwaitme

3e What kinds of manioc do you plant the most of in your garden?

4 Tú yujúmka imá pégkejaita nijamchísh

4e Which manioc makes the best manioc beer?

5 Tú yujúmka sénchi pushútuame

5e Which manioc makes a strong foam on manioc beer?

6 Wají yujúmka awí yúchataish, nijamchík takataísh

6e Which manioc do you not eat boiled, but only use for manioc beer?

7 Tú yujúmka imá pégkejaita awimásh

7e Which manioc is the best for boiling?

8 Tú yujúmka imá pégkejaita jiyamásh

8e Which manioc is the best for roasting?

9 Wají yujúmka udú yumaínaita

9e What kind of manioc can you eat raw?

10 Tú yujúmka apunásh nampuínaita

10e Which manioc grows the biggest roots?

11 Tú yujúmka imákuashtash iságmanash tsakáwa

11e Which manioc grows the tallest?

12 Wají yujúmka wámak nampuínaita

12e What kinds of manioc grow roots the most rapidly?

13 Wají yujúmka dipás nampuínaita

13e What kinds of manioc grow roots the most slowly?

14 Wají yujúmka néje katsújam wemaínchauwaita

14e What kinds of manioc roots do not go hard?

15 Wají yujúmka néje wámak katsújam wemaínaita

15e What kinds of manioc roots rapidly get hard?

16 Wají yujúmka imákuashtash chíchijaita

16e What kinds of manioc have the most amount of fiber in the root?

17 Wají yujúmka machikiúch chíchijaita

17e What kinds of manioc have the least amount of fiber?

18 Wají yujúmka wámak yapaú wemaínaita

18e What manioc can rapidly get bitter?

19 Wají yujúmka wámak jigkaíjuame

19e What kinds of manioc set fruit quickly?

20 Wají yujúmka dipás jigkaíjuame

20e What kinds of manioc are slow to set fruit?

21 Tú yujúmka wámak uwemaínaita

21e Which kinds of manioc can one harvest rapidly?

22 Tú yujúmka dipás uwemaínaita

22e Which kinds of manioc can one harvest slowly?

23 Tú yujúmka nujáhauame yúpichuchish

23e Which kinds of manioc are easy to peel?

24 Tú yujúmka tishimaínchau núja wéame

24e Which kinds of manioc are hard to peel?

25 Tú yujúmka awántak nejéame

25e Which kinds of manioc set their roots on the surface?

26 Tú yujúmka inítak nejéame

26e Which kinds of manioc set their roots deep in the ground?

27 Tú yujúmka uwítamush kaútsame

27e Which kinds of manioc do not rot after harvest?

28 Tú yujúmka uwítamush wámak kaúame
28e Which kinds of manioc rot rapidly after harvest?

29 Waj*í* yujúmka nujág amúgmash kaúchuita
29e What kinds of manioc do not rot after a flood?

30 Waj*í* yujúmka dáse umpuámash katúchauwaita
30e What kinds of manioc are not blown down by the wind?

The Ethnobotany of the Neotropical Solanaceae

Charles B. Heiser, Jr.

The family Solanaceae is well represented in the American tropics, and from here have come its greatest contributions to human welfare. Not surprisingly, the large genus *Solanum* has furnished the greatest number of them. Foremost among these is the Irish potato (*S. tuberosum* L.), which today ranks as one of the world's major food sources, exceeded in importance only by some members of the Gramineae. That this tetraploid species had its origin in the Andes is beyond doubt, but there has been no agreement as to its parentage. Recently, Hawkes and Cribb (in press) have postulated that *S. tuberosum* subsp. *andigenum* (Juz. & Buk.) Hawkes is derived from hybrids of the diploids *S. stenotomum* Juz. & Buk. and *S. sparsipilum* (Bitt.) Juz. & Buk., based on a resynthesis of the plant. Many other species of potato are still used for food in the Andes, largely to prepare "chuño," perhaps the original freeze-dried food. The potatoes are allowed to freeze at night, and as they thaw the next day the water is tramped out. After several days a completely dehydrated product results. There are two reasons for preparing "chuño": the process results in the removal of the bitter alkaloids present in wild potatoes (Brush, Carney & Huaman, 1981; Carlos Ochoa, pers. comm.), and the product can be stored indefinitely to be used as needed.

Not too distantly related to the potato, *S. muricatum* Ait., "pepino"—so named because the early Spanish fancied a resemblance of the fruits to the cucumber—is much appreciated for its fruits in the Andes. The plant, however, has not received much acceptance elsewhere. Three wild species, *S. basendopogon* Bitt., *S. caripense* H. & B. ex Dunal, and *S. tabanoënse* Corr., have been suggested as having figured in the origin of the "pepino" (Anderson, 1979, & unpubl.). One of these, *S. caripense,* is encountered in the medicinal markets in highland Ecuador. Its fruits are strung into necklaces and worn by small children to prevent "susto" (fright).

The section *Lasiocarpa* has also furnished a number of edible fruits. *Solanum quitoense* Lam., generally known as "naranjilla" in Ecuador and "lulo" in Colombia, clearly produces one of the finest fruits of any plant (Fig. 1a). Unfortunately, diseases are now plaguing the plant and fewer fruits reach the markets than did formerly, and these are quite expensive. The plant is generally grown at altitudes from 500 to 1500 m. A wild ancestor of the domesticate has not yet

FIG. 1.   a. "Naranjilla" (*Solanum quitoense*) near Baeza, Ecuador. b. "Tomate mora" (*Cyphomandra crassifolia*) near Patate, Ecuador.

been identified. A lowland species, *S. sessiliflorum* Dunal, known as "topiro," "cocono," or "cubiu," is cultivated in much of the Amazon basin for its fruits. The wild variety, *S. sessiliflorum* var. *georgicum* (R. E. Schultes) Whalen, has been identified as the probable progenitor of the domesticate. Several of the wild species in the section also have edible berries, and one of them, *S. pectinatum* Dunal, sometimes semi-cultivated by Indians, deserves a larger following, for its fruits also have an excellent flavor (Heiser, 1971; Whalen, Costich & Heiser, 1981).

One of the world's most important vegetables, the tomato (*Lycopersicon esculentum* Mill.), comes from the Neotropics. Although all of the wild species of *Lycopersicon* are native to South America, the domesticated tomato appears to have developed in Mexico (Rick, 1976).

Less well known is the tree tomato, *Cyphomandra crassifolia* (Ortega) Kuntze. Long appreciated in the Andes for its fruits, the plant is now being cultivated in many other parts of the world. In Ecuador, a new cultivar known as "tomate mora," much superior to the traditional "tomate de arbol," is now being grown (Fig. 1b). The new variety comes from New Zealand where it was developed from material from Loja, Ecuador.

The genus *Physalis* has provided two domesticated plants from the Neotropics. The "tomate" (*P. philadelphica* Lam.) was domesticated in Mexico from a wild variety of the species (Hudson, in press), and the fruits of both are widely used in Mexico and northern Central America for making green sauces. Such sauce is now available in the United States as green taco sauce. The berries of *P. peruviana* L. are eaten out of hand in the Andes, but the plants are seldom if ever cultivated there. The plant, however, was introduced to South Africa from Peru in the last century or earlier where it became cultivated. Later the plant went to Australia from Africa when it became known as Cape Gooseberry.

Four or five species of *Capsicum* were domesticated in tropical America (Eshbaugh, 1980; Heiser, 1976; Pickersgill, Heiser & McNeill, 1979). The fruits furnish both food and spice, and next to black pepper (*Piper nigrum* L.), chili peppers are the world's most important spice. These peppers were introduced to Europe following Columbus' voyages and are now widely grown throughout the world, particularly in the tropics. The extremely pungent wild pepper, *C. annuum* var. *glabriusculum* (Dunal) Heiser & Pickersgill, is also widely used as a spice in the southwestern United States, Mexico, and Guatemala, as are some of the wild species in South America.

The family Solanaceae includes a number of genera that have been used as narcotics, particularly as hallucinogens, in the Americas. Much of our knowledge of these comes from the work of Schultes (1979) and his students. Although it may seem strange to include tobacco (*Nicotiana tabacum* L.) in a list of plants that has contributed to human welfare, the plant is clearly still one of the most important economically, and taxes derived from it have sometimes served worthy causes. A second narcotic species (*N. rustica* L.), now little used, was under cultivation in Mexico and eastern North America at the time of the Discovery. Cytogenetic evidence indicates that both species had their origin in South America (Gerstel, 1976). Most prominent among the hallucinogens are the tree Daturas, several species of which are so employed (Lockwood, 1979). Seeds of one of these, *Brugmansia sanguinea* (R. & P.) D. Don (Fig. 2), are still added to maize "chicha" in parts of the Andes to give extra strength to the beverage. This species has also recently been brought into cultivation in Ecuador for the commercial production of scopolomine from the leaves.

Another solanaceous species used in medicine is *Solanum marginatum* L.f. Although native to Africa, this species has been well established as a weed in the

Fig. 2.   The tree Datura (*Brugmansia sanguinea*) near Patate, Ecuador.

northern Andes for some time where its fruits have been used as a substitute for soap for washing clothes. A few years ago, the late Dr. Alfredo Paredes of the Escuela Politecnica Nacional of Ecuador found that its fruits were a rich source of steroids, and it is now being grown in Ecuador for the commercial production of solasodine (Roth, in press) which goes to West Germany for the production of anti-inflammatory drugs and birth control pills.

Finally, the family has contributed a large number of ornamentals (Bailey, 1949). Most of these come from South America. Several of the better-known ones, such as *Petunia hybrida* Vilm., *Salpiglossis sinuata* R. & P., and *Schizanthus pinnatus* R. & P., come from Chile or Argentina; but many of the others, including representatives of several of the genera already discussed, come from the tropics. The shrub *Streptosolen jamesonii* Miers, although not well adapted to the United States, is widely cultivated in western South America were, in some places, it is known under the picturesque name "lluvia de estrellas" (shower of stars). Among the recent introductions to horticulture is the widespread tropical weed *Solanum mammosum* L. The fruits have long been used in Guatemala in domestic medicine and as adornment by women on the pilgrimage to the Sanctuary of Esquipulas— apparently in the belief that it will result in childbirth (Gentry & Standley, 1974). In Ecuador they are used for killing cockroaches. Today the plant is grown for the branches with mature fruits which are used in dried winter boquets. Apparently, the Japanese were the first to grow it for this purpose, and seeds are now offered by a company in the United States.

Acknowledgments

I acknowledge Luis Levy, Carlos Ochoa, Leo Roth, and Sergio Soria for providing information on some of the plants discussed above. This paper is a summary of one presented at the Economic Botany symposium. Many of the details given in the oral presentation, but lacking here, can be found in Heiser (1969).

Literature Cited

Anderson, G. J. 1979. Systematic and evolutionary consideration of species of *Solanum*, section *Basarthrum*. Pages 549–562. *In:* J. G. Hawkes, R. N. Lester & A. D. Skelding, editors. The biology and taxonomy of the Solanaceae. Academic Press, New York.

Bailey, L. H. 1949. Manual of cultivated plants. MacMillan, New York.

Brush, S. B., H. J. Carney & Z. Huaman. 1981. Dynamics of Andean potato agriculture. Econ. Bot. 35: 70–88.

Eshbaugh, W. H. 1980. The taxonomy of the genus *Capsicum* (Solanaceae)—1980. Phytologia 47: 153–166.

Gentry, J. L. & P. C. Standley. 1974. Solanaceae. *In:* Flora of Guatemala. Fieldiana 24: 1–151.

Gerstel, D. U. 1976. Tobacco. Pages 273–277. *In:* N. W. Simmonds, editor. Evolution of crop plants. Longman, London.

Hawkes, J. G. & P. Cribb. In press. Experimental evidence for the origin of *Solanum tuberosum* subsp. *andigenum* (Juz. et Buk.) Hawkes. *In:* W. G. D'Arcy, editor. Biology and systematics of the Solanaceae. Columbia University Press, New York.

Heiser, C. B. 1969. Nightshades, the paradoxical plants. W. H. Freeman, San Francisco.

———. 1971. Notes on some species of *Solanum* (Sect. *Leptostemonum*) in Latin America. Baileya 18: 59–65.

———. 1976. Peppers. Pages 265–268. *In:* N. W. Simmonds, editor. Evolution of crop plants. Longman, London.

Hudson, W. D. In press. The relationship of domesticated and wild *Physalis philadelphica*. *In:* W. G. D'Arcy, editor. Biology and systematics of the Solanaceae. Columbia University Press, New York.

Lockwood, T. E. 1979. The ethobotany of *Brugmansia*. Journ. Ethnopharm. 1: 147–164.

Pickersgill, B., C. B. Heiser & J. McNeill. 1979. Numerical taxonomic studies of variation and domestication in some species of *Capsicum*. Pages 679–700. *In:* J. G. Hawkes, R. N. Lester & A. D. Skelding, editors. The biology and taxonomy of the Solanaceae. Academic Press, New York.

Rick, C. M. 1976. Tomato. Pages 268–273. *In:* N. W. Simmonds, editor. Evolution of crop plants. Longman, London.

Roth, L. In press. World-wide distribution of the wild plant *Solanum marginatum* and its cultivation in Ecuador. *In:* W. G. D'Arcy, editor. Biology and systematics of the Solanaceae. Columbia University Press, New York.

Schultes, R. E. 1979. Solanaceous hallucinogens and their role in the development of New World cultures. Pages 137–160. *In:* J. G. Hawkes, R. N. Lester & A. D. Skelding, editors. The biology and taxonomy of the Solanaceae. Academic Press, New York.

Whelan, M. D., D. E. Costich & C. B. Heiser. 1981. Taxonomy of *Solanum* section *Lasiocarpa*. Gentes Herbarum 12: 41–129.

Plants and Dental Care among the Jívaro of the Upper Amazon Basin

Walter H. Lewis and Memory P. F. Elvin-Lewis

In the Upper Amazon Basin, the Jívaro are among the most numerous and most cultured Amerindians. By successfully thwarting most acculturation, the Jívaro have retained much of their heritage (Lockwood, 1979), including a wide pharmacological knowledge in which plants play a dominant role (Karsten, 1935). Modern medicine clearly owes a number of useful plants to the Jívaro and other tribes in and around the Amazon Basin, such as *Chondrodendron tomentosum* Ruíz & Pavón (*d*-tubocurarine from curare), but according to Karsten "those hitherto known only form a fraction of those actually existing. I therefore wish to emphasize once more the importance of this particular subject being studied by experts in pharmacology and botany. Expeditions sent out for this purpose to the upper Amazonas regions are almost sure to achieve results of great value."

In the Old World, dental health involving low caries rates and minimal teeth loss is correlated with the widespread use of chewing sticks and sponges (Elvin-Lewis et al., 1980a; Elvin-Lewis, 1982). In the neotropics, however, chewing sticks are used only occasionally in the Caribbean Basin area, while in the vast region of the Amazon and other parts of South America, this means of dental care is rare or absent. What then are the alternative methods of folk dentistry? To answer this intriguing question, we travelled to the Upper Amazon Basin of Peru in 1982 where we studied the dental practices of two Jívaro groups; the Achual on the Ríos Huasaga, Pastaza, and Tigre, and the Mayna on the Río Corrientes (Fig. 1).

Teeth care by blackening

Karsten (1935) reported two plants, "nashumbi" and "píu," used by the Jívaro to blacken their teeth, but he did not know why they did so. However, Clark (1954), who also observed that "nushumbi" was used by the Jívaro to paint their teeth black, reported that this practice was said to prevent tooth decay. Clark's "nushumbi" and Karsten's "nashumbi" may have been *Calatola costaricensis* which we found to be used for teeth blackening among the Mayna who called it "nashum." The "píu" of Karsten was either *Calatola* or a species of *Neea*, the

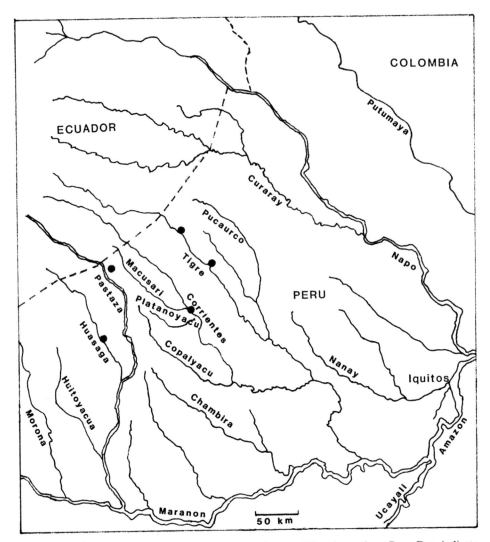

FIG. 1.   Map showing the major tributaries of the Amazon River in northern Peru. Dots indicate the approximate localities of the major communities visited: Washi Intza (Huasaga), Tintiyacu (Pastaza), Pampa Hermosa (Corrientes), Marsella (upper Tigre), and Vista Alegra (lower Tigre).

leaves of which we found to be a common source of teeth blackening used by the Achual. In Colombia, Klug (Standley, 1936) observed that leaves of *Neea parviflora* ("yana muco") were chewed by the Indians to blacken their teeth, which he noted were very sound and strong. These and other parts of species of *Calathea, Dendropanax, Duroia, Manettia, Petiveria, Piper, Schradera,* and *Siphocampylus* used for teeth blackening in northwestern South America are listed in Table I.

Teeth blackening by leaf mastication to prevent tooth decay may be unique to populations in South America and adjacent Central America, but by other means and for cosmetic purposes, blackening of teeth was once widely practiced in southeastern Asia and Japan (Ai et al., 1965; reviewed by Lewis & Elvin-Lewis, 1977, pp. 266–269). In that region, plant tannins from bark, acorns, nutgalls, ashes, tar, stems, etc., were used alone or in combination with iron to obtain blackening for prolonged periods.

One can only speculate on the substances in the leaves of these different plants and the mechanisms by which they prevent tooth decay. Because phenols or their polymers usually oxidize and turn black on exposure to air and because under the basic conditions found in the mouth this process would be hastened, compounds such as 1,2-quinone or tannin could be responsible for the rapid and prolonged blackening observed during mastication of fresh leaves. Several commercially important species of *Piper,* for example, have prominent phenolic compounds, such as eugenol and its isomers, and tannin in leaves (Krishnamurthi, 1969). Tannins from tea (Elvin-Lewis et al., 1980b) and cocoa (Paolino et al., 1980) are known to affect the cariogenic process by inhibiting enzymes associated with bacterial adherence. It is not known whether the substances in teeth-blackening plants act similarly or whether they serve as sealants by coagulating the plaque protein and stabilizing the hydroxyapatite of teeth (Ai et al., 1965).

Nevertheless, it was the impression of dental pathologists at Washington University School of Dental Medicine, who examined intraoral photographs of habitual teeth blackeners and of others who no longer practice the habit, that the dental health among the former group was better than among the latter. For example, with the exception of limited gingivitis, which might be expected if calculus had formed, no other dental disease was apparent among the habitual teeth blackeners (Fig. 2a). In the other group, and also among Peruvians of European origin, proximal caries, especially in the upper maxillary teeth, were frequently seen, and periodontal disease associated sometimes with poor gold crowns was also observed. It should be emphasized that other forms of oral hygiene, such as tooth brushing, are not routine among the Jívaro and that no other preventive measures are practiced systematically.

Teeth reddening

At times young Mayna women masticated bark of *Simira rubescens* (Benth.) Bremek. ex Steyerm. (Rubiaceae) to stain their oral cavity and particularly their teeth red. This appears to be a cultural practice unrelated to hygiene which is used to enhance beauty and attractiveness. This practice was not observed among Achual females.

Teeth extraction

The Jívaro today who develop diseased and aching teeth from the lack of dental hygiene apparently do not use toothache remedies *per se* but do extract teeth that ache. When a tooth becomes painful because of traumatic injury, malocclusion, periodontal disease, or other reasons, the tooth is removed by using the sap or latex of a number of plants that also cause the pain to rapidly disappear, presumably by acting as an analgesic. A piece of kapok is usually soaked in the sap or latex and placed in the vicinity of the pain. The tooth disintegrates and falls out in pieces or is removed with little trauma or blood within a week or more. The technique is practiced largely by women, particularly those of childbearing age. None of the men interviewed knew how to extract teeth or even which plants were used; presumably Jívaro women alone are the custodians of this indigenous dental practice. We observed that some men had dental restorations, especially partial gold crowns of the maxillary incisors. It was often these teeth that were missing among the women.

Table I

Teeth-blackening plants and their uses

Family	Species	Vernacular name	Use			Reference or herbarium voucher
			Plant part	Locality	Indian group	
Araliaceae	*Dendropanax tessmannii* (Harms) Harms	cheriz	leaves	Peru (Río Putumayo)		*Klug 2016* (MO)
Campanulaceae	*Siphocampylus giganteus* (Cav.) G. Don		latex	Colombia		García-Barriga (1975)
Icacinaceae	*Calatola costaricensis* Standley	pío	leaves	Peru (Río Pastaza)	Achual	*Lewis et al. 4008* (MO)
		nashum	leaves	Peru (Río Corrientes)	Mayna	*Lewis et al. 4112* (MO)
Marantaceae	*Calathea* sp.		seeds	Peru (Amazonas)	Aguaruna	B. Berlin (pers. comm.)
Nyctaginaceae	*Neea* cf. *floribunda* P. & E.	yanamuco	leaves	Peru (Río Tigre)	Achual	*Lewis et al. 4137* (MO)
	N. parviflora P. & E.	yana muco	leaves	Colombia	Cacueta, Putumayo	Standley (1936)
	N. cf. *divaricata* P. & E.	píocha	leaves	Peru (Río Pastaza)	Achual	*Lewis et al. 4007* (MO)
Phytolaccaceae	*Petiveria alliacea* L.	anamu	plant	Colombia		García-Barriga (1974)
Piperaceae	*Piper hispidum* Sw.	ungushurat	leaves	Peru (Río Huasaga)	Achual	*Lewis et al. 4101* (MO)
	P. marginatum Jacq.		plant	Colombia	Katio	García-Barriga (1974)
	P. tingens Trelease	cordoncillo	plant	Peru (Río Paranapura)		Altschul (1973)
Rubiaceae	*Duroia hirsuta* (P. & E.) K. Sch.		fruit	Colombia		A. Fernandez (pers. comm.)

Table I
Continued

Family	Species	Vernacular name	Plant part	Use		Reference or herbarium voucher
				Locality	Indian group	
	Manettia divaricata Wernham	yanamuco	leaves			Ayala 575 (AMAZ), Woytkowski 1527 (MO), Altschul (1973)
	M. glandulosa P. & E.			Peru (Pucallpa)		Woytkowski 5771 (MO)
	Schradera marginalis Standley	queda, quera	plant	Colombia	Citará	Archer (1934)
Undetermined		tzu'-he	leaves	Colombia		Schultes 3508 (GH)
		yotoconti		Ecuador? (Oriente)	Campa	Clark (1954)

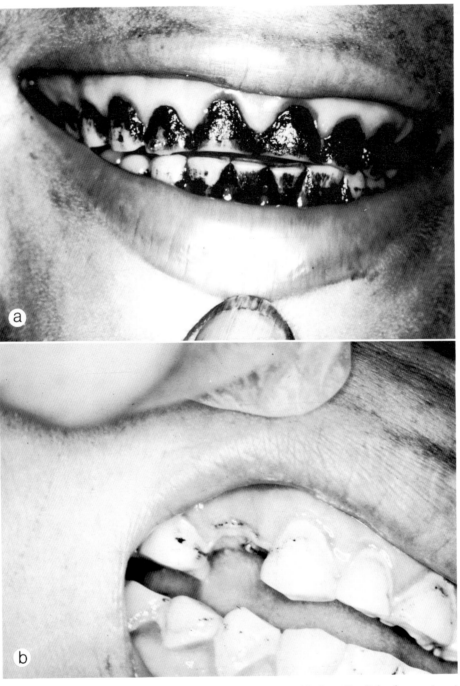

FIG. 2.   a. A traditional Mayna Jívaro female's mouth with teeth blackened by *Calatola costaricensis* leaves. b. An Achual Jívaro female's mouth showing the absence of a tooth extracted by using the pulp of immature pericarps of *Genipa americana*.

Table II
Teeth-extracting plants and their use

Family	Species	Vernacular name	Plant	Use Locality	Indian group	Reference or herbarium voucher
Asclepiadaceae	*Asclepias curassavica* L.		latex	Colombia		Pérez-Arbeláez (1975)
Malpighiaceae	*Stigmaphyllon* sp.		stem sap	Peru (Río Tigre)	Achual	*Lewis et al. 4134* (MO)
Moraceae	*Chlorophora tinctoria* (L.) Gaud.	dinde	latex	Colombia (Cundinamarca)	country people	*Lewis & Elvin-Lewis 9784-9486* (MO), García-Barriga (1974)
		incira		Peru (Río Algodón)	Coto	Maxwell (1962)
				Panama	Choco	Lewis & Elvin-Lewis (1977)
	Coussapoa glaberrima W. Burger	sararay-wa-remba-yed, cani	fruit	Venezuela		Reis & Lipp (1982)
Rubiaceae	*Genipa americana* L.	sua	immature pericarp	Peru (Río Pastaza)	Achual	*Lewis et al. 4004* (MO)

The latex of *Chlorophora tinctoria* (Moraceae) has been used to extract teeth in Peru by the Coto along the Río Algodon (Maxwell, 1962), in Colombia (García-Barriga, 1974; Pérez-Arbeláez, 1975), and in Panama (Lewis & Elvin-Lewis, 1977). It has been reported that *Asclepias curassavica* (Asclepiadaceae) (Pérez-Arbeláez, 1975) and the fruit of *Coussapoa glaberrima* (Moraceae) (Reis & Lipp, 1982) are used in Colombia and Venezuela, respectively, for extracting teeth. We found that among the Achual both the stem sap of *Stigmaphyllon* sp. (Malpighiaceae) and the pulp of the immature pericarp of *Genipa americana* (Rubiaceae) are used also for this purpose (Table II). According to anecdotal information, the techniques appear to be effective and healing to be normal. We observed, however, a root-stump remaining in a user of *G. americana,* suggesting that incomplete extraction can occur if the application is not done correctly (Fig. 2b). Moreover, when an excessive amount of the plant material is used, it may also cause the removal of healthy, adjacent teeth.

Studies of the latex of *C. tinctoria* have been limited to a few preliminary experiments. It does not possess collagenase activity, but its low pH indicates that the latex may be proteolytic in nature. That the proteolytic enzyme ficin has been isolated from *Ficus* (also in the Moraceae) and that the latex from fig has also been used to treat toothaches (Elvin-Lewis, in press) suggest that a similar compound may be involved in the case of *C. tinctoria.* Although limited data for *G. americana* show that its fruit contains a highly oxygenated cyclopentanoid monoterpene called genipin (Djerassi et al., 1961) with broad spectrum antibacterial action (Cordova-Marquez et al., 1954), compounds responsible for its analgesic and/or tooth-disintegrating action have not yet been identified.

Summary

As the acculturation of Jívaro communities continues, the knowledge and use of teeth blackening for prevention of caries are disappearing. Yet alternative dental methods are not routinely practiced in the communities which we visited and are only available to those who have the funds and can seek dental treatment in far-removed urban areas. No attempt is made, for example, to teach school children preventative methods of dental hygiene, such as tooth brushing, and even if brushes were made available, toothpastes or other substances needed for plaque removal (e.g., salt, wood ashes) are unavailable or unknown. Currently, only the more traditional Jívaro believe that tooth blackening prevents tooth decay and, hence, toothache and tooth loss, and, therefore, continue this practice. The major reason for the abatement of this practice appears to be the Jívaro's sensitivity to the non-Indian aversion to blackened teeth. The Indians are thus discarding their only means of preventative dentistry in much the same way as the Japanese, in the late 19th century, stopped blackening their teeth in response to Western repugnance toward it. Nevertheless, resumption of the practice appears to be the only viable alternative for those Jívaro who remain in their communities where a supply of fresh leaves are obtainable without cost. This practice should be strongly recommended if research substantiates the efficacy of teeth blackening in preventing dental caries. Furthermore, possible applications of teeth-extraction methods used by the Jívaro to modern oral surgery and orthodontics need study.

Acknowledgments

The expedition was sponsored by a grant (2510-82) from the National Geographic Society, and further assistance in the field was contributed by the Occidental Exploration and Production Company, Bakersfield, California. We thank

Dr. Franklin Ayala, Rodolfo Vásquez, Nestor Jaramillo, and José Torres of the Universidad Nacional de la Amazonia Peruana, Iquitos; Carlos Grandez, Iquitos; Eduardo Gonzalez B., Marsella; and Dr. José Perea S., Universidad Nacional de Colombia, Bogotá. We also thank Dr. Arthur Eisen, Department of Dermatology, Washington University School of Medicine, for examining *Chlorophora tinctoria* for collagenase activity, and Drs. Charles Waldron, Shirley Pierce, and Samir El-Mofty, Washington University School of Dental Medicine, for studying the intraoral photographs of Achual and Mayna Jívaro.

Literature Cited

Ai, S., T. Ishikawa & A. Seino. 1965. "Ohaguro" traditional tooth staining custom in Japan. Intern. Dent. J. 15: 426–441.

Altschul, S. von Reis. 1973. Drugs and foods from little-known plants: Notes in Harvard University Herbaria. Harvard University Press, Cambridge.

Archer, W. A. 1934. The dental plant of the Citará Indians in Colombia. J. Washington Acad. Sci. 24: 402–404.

Clark, L. 1954. The rivers ran east. Hutchinson, London.

Cordova-Marquez, R., J. H. Axtmayer & A. Brenes-Pomales. 1954. Antibacterial action of an extract from the jagua fruit (*Genipa americana*). Bol. Asoc. Med. Puerto Rico 46: 375–376.

Djerassi, C., T. Nakano, A. N. James, L. H. Zalkow, E. J. Eisenbraun & J. N. Schoolery. 1961. Terpenoides XLVII. The structure of genipin. J. Org. Chem. 26: 1192–1206.

Elvin-Lewis, M. P. F. 1982. The therapeutic potential of plants used in dental folk medicine. Odontostomatol. Trop. 5: 107–117.

———. In press. Therapeutic rationale of plants used to treat dental infections. *In:* N. L. Etkin, editor. Plants used in indigenous medicine and diet. Redgrave Publ., South Salem, New York.

———, **J. B. Hall, M. Adu-Tutu, Y. Afful, K. Asante-Appiah & D. Lieberman.** 1980a. The dental health of chewing-stick users of southern Ghana: Preliminary findings. J. Prev. Dent. 6: 151–159.

———, **M. Vitale & T. Kopjas.** 1980b. Anticariogenic potential of commercial teas. J. Prev. Dent. 6: 273–284.

Garcia-Barriga, H. 1974–75. Flora medicinal de Colombia—botánica médica, Vol. I–II. Instituto de Ciencias Naturales Universidad Nacional, Bogotá.

Karsten, R. 1935. The head-hunters of western Amazonas. Com. Hum. Lit., Soc. Sci. Fennica 7: 1–598.

Krishnamurthi, A., editor. 1969. The wealth of India, Vol. 8. Publications & Information Directorate, C.S.I.R., New Delhi.

Lewis, W. H. & M. P. F. Elvin-Lewis. 1977. Medical botany: Plants affecting man's health. Wiley-Interscience, New York.

Lockwood, T. E. 1979. The ethnobotany of *Brugmansia*. J. Ethnopharmacol. 1: 147–164.

Maxwell, N. 1962. Witch doctor's apprentice. Victor Gollancz, London.

Paolino, V. J., S. Kashket & C. A. Sparagna. 1980. Inhibition of dextran synthesis of tannic acid. J. Dent. Res. 59: 389.

Pérez-Arbeláez, E. 1975. Plantas medicinales y venenosas de Colombia. Hernando Salazar, Medellín, Colombia.

Reis, S. von & F. J. Lipp, Jr. 1982. New plant sources for drugs and foods from the New York Botanical Garden Herbarium. Harvard University Press, Cambridge, Massachusetts.

Standley, P. C. 1936. Studies of American plants—VI. Nyctaginaceae. Field Mus. Nat. Hist., Bot. Ser. 11: 153–154.

The Ethnobotany of Coca (Erythroxylum spp., Erythroxylaceae)

Timothy Plowman

Of all the botanical wonders discovered in the New World by the first European explorers, few can compare with the coca plant for its fascinating history, its remarkable medicinal properties, and its continuing economic and political importance. For millions of South American natives, coca not only furnishes a mild stimulant and sustenance for working under harsh environmental conditions, but also serves as a universal and effective household remedy for a wide range of medical problems. The traditional use of coca also plays a crucial symbolic and religious role in Andean society. Its use is accompanied by complex rituals, ceremony and protocol, such that coca functions as a focus of cultural and social integration. It has been said that chewing coca is the most profound expression of Andean culture and that, if coca were taken away from the Indians, their traditional culture would rapidly disintegrate (Wagner, 1978; Carter et al., 1980a, 1980b).

In sharp contrast to the unifying and stabilizing effects of coca chewing on Andean culture is the disruptive and convoluted phenomenon of cocaine use in Western societies. Because all cocaine entering world markets is derived from coca leaves produced in South America, the staggering increase in demand for cocaine for recreational use has a direct impact on South American economies, politics and, ultimately and most tragically, on indigenous cultures. The widespread use of cocaine, either for pleasure or work, is a very different psychological experience than using coca in a traditional setting; the differences between the pharmacological effects of cocaine hydrochloride taken, say, intranasally, and the effects of chewing coca leaves have been emphasized repeatedly (Mortimer, 1901; Weil, 1975; Grinspoon & Bakalar, 1976; Antonil, 1978; inter alia). Yet many people still equate the use of coca with that of cocaine and fail to comprehend both the pharmacological and cultural differences between these two related yet unique substances. In modern societies, people are fairly well acquainted with both the pleasurable and deleterious effects of cocaine because of extensive news coverage of the cocaine "phenomenon" in recent years. Yet few people are aware of the beneficial effects of coca chewing, of the importance of the use of coca in Andean life, or of the origin and evolution of the coca plant.

During the past decade, we have seen enormous progress in research on the history, chemistry, botany, and cultural importance of coca. Unfortunately, most of these studies have been overshadowed by a much greater profusion of studies on the pharmacology and chemistry of cocaine and on its physiological and psychological effects. It is my purpose here to present an overview of the botany, chemistry, and uses of coca by South American natives and to review pertinent research on coca which has appeared since approximately 1970. Areas of particular interest include recent studies on the botanical origins of coca, which until the 1970's remained muddled and misunderstood even by taxonomic botanists; on the archeological record of coca, which, although rather scanty, had been largely misinterpreted by archeologists; and on the chemistry of the coca leaf, which had never been adequately analyzed because of earlier technical problems in making efficient extractions and quantitative measurements of the contained compounds. There has also been renewed interest in the effects of coca chewing, but we still know relatively little about the subtle and complex pharmacology of the experience. Lastly, there has been an effort on the part of anthropologists to document more completely the religious and cultural aspects of coca in traditional cultures.

One area of study which I will not consider here is the long and complex history of coca after the Spanish Conquest. This topic has been investigated in depth by many scholars over a long period of time but space limitations preclude discussing it here. The reader is referred to the following works in which new and noteworthy findings on the history of coca during the Colonial period are presented: Uscátegui, 1954; Gagliano, 1960, 1963, 1965, 1968, 1979; Patiño, 1967; Martin, 1970; Peña Begué, 1972; Burchard, 1976; Chávez Velásquez, 1977; Antonil, 1978; Carter et al., 1980a; Castro de la Mata, 1981; Bray & Dollery, 1983; Plowman, 1984).

Botany of coca

Many scholars have underestimated or overlooked entirely the importance of the existence of distinct varieties of coca. Although geographical, ecological, and morphological differences in coca varieties were recorded as early as the 16th century, their significance was not recognized until the 1970's (Rostworowski, 1973; Antonil, 1978; Plowman, 1979a; Bray & Dollery, 1983). Not until coca leaf became an important pharmaceutical product in the late 19th century, did the botanical origins and varieties of coca become the object of scientific inquiry (Plowman, 1982).

The coca shrub belongs to the genus *Erythroxylum* P. Browne of the tropical plant family Erythroxylaceae. Most species of *Erythroxylum* are found in the American tropics with about 200 species, although the genus also occurs in Africa, Madagascar, India, tropical Asia, and Oceania. In the Old World, many wild species are employed in folk medicine (Hegnauer, 1981), but it is only in tropical America where *Erythroxylum* leaves are chewed extensively as a stimulant and where the plants attain major cultural importance (Martin, 1970; Mayer, 1978; Antonil, 1978; Carter et al., 1980a).

All cultivated coca is derived from two closely related South American species— *Erythroxylum coca* Lam. and *E. novogranatense* (Morris) Hieron. Whereas other neotropical wild species of *Erythroxylum* may be employed locally as medicines, discussions of "coca" should be confined to these two species.

Until relatively recently, only one species of coca—*Erythroxylum coca*—was generally recognized (Mortimer, 1901; Hegnauer & Fikenscher, 1960; Martin,

1970). However, evidence, resulting from intense field and laboratory studies of coca, has accumulated during the past decade and demonstrates incontrovertibly that two distinct species of coca should be recognized (Schulz, 1907; Machado, 1972; Gentner, 1972; Plowman, 1979a; Rury, 1981, 1982; Bohm et al., 1982; Plowman & Rivier, 1983). In addition, each of the two species of cultivated coca has one variety, designated *E. coca* var. *ipadu* Plowman and *E. novogranatense* var. *truxillense* (Rusby) Plowman, respectively. The four cultivated cocas of South America are thus treated as follows: *E. coca* var. *coca, E. coca* var. *ipadu, E. novogranatense* var. *novogranatense,* and *E. novogranatense* var. *truxillense.*

All of the varieties of cultivated coca were domesticated independently in pre-Columbian times and are still employed by native coca chewers in South America. Each of them was known by a different native name before the Spanish popularized the now widespread term "coca." Although they differ appreciably in the content of minor alkaloids and other chemical constituents, all of the cultivated cocas contain the alkaloid cocaine. Additional important differences among the four varieties, which hitherto have been overlooked, are found in their leaf and stem anatomy, ecology, geographical relationships, and in the methods of their cultivation and preparation for chewing. These differences reflect intensive human selection over a long period of time for specific traits and for successful cultivation in a variety of habitats in distinct geographic areas (Fig. 1).

Although certain wild species may yet be implicated in their evolutionary relationships (Plowman & Rivier, 1983), the four varieties of cultivated coca are more closely related to each other than to any other species of *Erythroxylum.* Superficially, the cultivated cocas are very similar morphologically, which explains in part earlier confusion in the identification of coca specimens, especially by non-specialists (Plowman, 1979b, 1982). The varieties can be distinguished by characters of the branching habit, bark, leaves, stipules, flowers, and fruits; but often, especially in the case of dried herbarium specimens, complete specimens may be necessary for positive identification. However, in most cases isolated coca leaves can now be identified to species if not to variety, especially if the provenance of the samples is known.

Recent studies have provided additional new characters that permit the accurate and positive identification of coca leaves, including archeological specimens. These studies focus on leaf anatomy (Rury, 1981, 1982; Rury & Plowman, 1984), flavonoids (Bohm et al., 1981), alkaloids (Rivier, 1981; Plowman & Rivier, 1983), reproductive biology and breeding relationships (Ganders, 1979; Bohm et al., 1982), and ecology and geographic distribution (Plowman, 1979a, 1979b, 1984). As a result of these investigations, the taxonomic and evolutionary relationships among the four cultivated cocas are now fairly well understood.

Erythroxylum coca var. coca, Huánuco or Bolivian coca

Erythroxylum coca consists of the wide-ranging and economically important Andean variety of *E. coca* var. *coca* and the geographically restricted Amazonian variety *E. coca* var. *ipadu. Erythroxylum coca* var. *coca* is often referred to as "Bolivian" or "Huánuco" coca, but neither of these terms conveys the extensive geographical range of the variety. For convenience, I will use the term "Huánuco coca" here.

Erythroxylum coca var. *coca,* a shrub 1 to 3 m tall (Fig. 2), grows mainly between 500 and 1500 m elevation but may reach 2000 m in some areas. It is cultivated in regions of moist, montane tropical forest along the eastern slopes of the Andes and in the wetter inter-Andean valleys, in the ecological zone known generally as

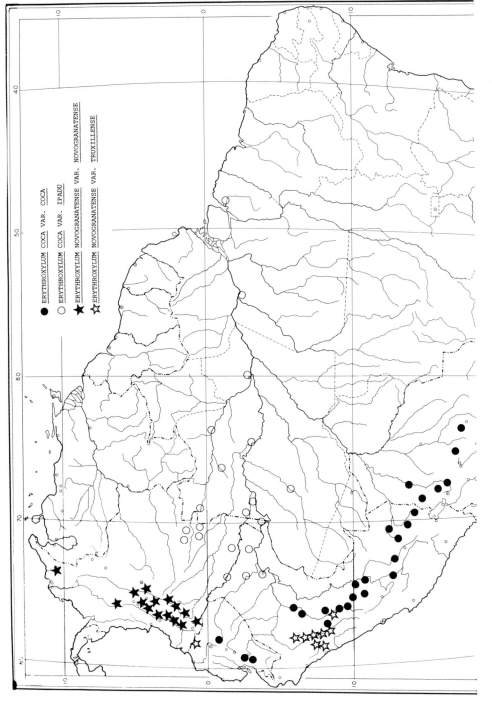

FIG. 1. Present distribution of the four varieties of cultivated coca (*Erythroxylum* spp.) based on herbarium collections.

FIG. 2.   Habit of a mature, fruiting specimen of *Erythroxylum coca* var. *coca* cultivated at Tarapoto, Dept. San Martín, Peru (*Plowman 6042*).

"montaña" (Fig. 3). Because it has a fairly limited ecological range, Huánuco coca is little known outside its original area in South America. However, this variety is the principal commercial source of coca leaves and of most of the world's cocaine supply.

Fig. 3.   Plantation of *Erythroxylum coca* var. *coca* in the moist, tropical montaña habitat, San Francisco, Río Apurímac, Dept. Ayacucho, Peru (*Plowman & Jacobs 4711*).

The geographical distribution of Huánuco coca extends from Ecuador south to Bolivia and northwesternmost Argentina (Fig. 1). Only in Ecuador, where suitable moist forest habitats occur on both sides of the Andes, does this variety reach the Pacific slope. It is unknown in Colombia or in the Amazon lowlands.

Throughout its range, Huánuco coca is found as wild-growing or feral individuals in the understory of primary or secondary forests, both near and remote from areas of present coca cultivation. It is well adapted to the montaña habitat where it appears to be a natural component of the forest understory and occurs sympatrically with several wild erythroxylums including *E. ulei* O. E. Schulz, *E. mamacoca* Mart., *E. macrocnemium* Mart., and *E. mucronatum* Benth.

It is often impossible to distinguish between truly wild-growing *E. coca* var. *coca* and plants that have escaped from coca plantations or that persist after plantations are abandoned. There are apparently no barriers to gene flow between wild and cultivated populations, which freely interbreed when growing in proximity. The small red fruits are eaten by birds which disseminate the seeds throughout the montaña habitat. There are no essential structural differences between wild-growing and cultivated plants of *E. coca* var. *coca,* and this variety seems to be little altered morphologically, genetically, or physiologically through domestication. In this feature, *E. coca* var. *coca* differs fundamentally from many other cultivated plants, especially food plants, which may become isolated genetically from their wild progenitors and lose their ability to reproduce in the wild (Pickersgill & Heiser, 1976).

Erythroxylum coca var. *coca* is now thought to be a naturally occurring wild species of the montaña, from which the other three cocas ultimately were derived as cultigens through human selection. Probably *E. coca* var. *coca* had a more limited distribution as a wild species, possibly in eastern Peru in the area centering on the Huallaga Valley, where it frequently is found growing wild. Subsequent range extensions northward to Ecuador and southward to Bolivia and Argentina probably occurred through man's activities.

ERYTHROXYLUM COCA VAR. IPADU, AMAZONIAN COCA

Although long-neglected by anthropologists, Amazonian coca, *Erythroxylum coca* var. *ipadu,* recently has been re-examined by botanists (Prance, 1972; Plowman, 1979b, 1981; Schultes, 1981; Rury, 1981, 1982; Plowman & Rivier, 1983) and pharmacologists (Holmstedt et al., 1979). Amazonian coca is closely allied to *E. coca* var. *coca* from which it has originated in relatively recent times (Plowman, 1981). The Amazonian variety is cultivated on a small scale by a number of tribes of the upper Amazon in parts of Colombia, Brazil and Peru (Fig. 1). Propagated by stem cuttings, it is well adapted to the pattern of shifting agriculture practiced by semi-nomadic Amazonian peoples. Amazonian coca does not survive as a feral or escaped plant in the lowland Amazon and may be considered a true cultigen.

Amazonian coca is little differentiated from *E. coca* var. *coca,* and the two varieties appear to be fully interfertile. Amazonian coca contains the same leaf flavonoid profiles as the montaña variety. The leaf flavonoids have been found to be a useful and unvarying taxonomic character for identifying both cultivated and wild cocas (Bohm et al., 1981, 1982). A surprising chemical difference in Amazonian coca is a consistently lower cocaine content; this variety usually contains only about half the concentrations found in other cultivated cocas (Holmstedt et al., 1977, 1979; Plowman & Rivier, 1983).

Erythroxylum coca var. *ipadu* was unknown to Europeans until the middle of the 18th century. Details of its cultivation, use, and geographic distribution were not recorded until the present century. Amazonian coca has no archeological record with which to date its origin in Amazonia, but based on linguistic, ethnographic, historical and botanical evidence, Amazonian coca appears to be a relatively recent development. It surely evolved from stocks of *E. coca* var. *coca* introduced from the Andean foothills through selection for traits conducive to its cultivation in Amazonia. It is now geographically isolated from other coca varieties.

ERYTHROXYLUM NOVOGRANATENSE

Erythroxylum novogranatense now is recognized as a distinct species of cultivated coca, although in the past it often was confused with, or considered a variety of, *E. coca* (Plowman, 1982). Appreciable evidence has accumulated that suggests that this species arose as a domesticated plant through human selection from *E. coca* var. *coca* (Bohm et al., 1982). *Erythroxylum novogranatense* differs from *E. coca* var. *coca* in a number of morphological features, but more significantly, it has evolved distinctive chemical and ecological traits and has become genetically isolated from parental *E. coca* var. *coca.*

Erythroxylum novogranatense consists of two well defined varieties: *E. novogranatense* var. *truxillense,* Trujillo coca, and *E. novogranatense* var. *novogranatense,* Colombian coca. These varieties are more strongly differentiated from each other than *E. coca* var. *coca* is from *E. coca* var. *ipadu.* This suggests greater antiquity for the varietal isolation and differentiation within *E. novogranatense* than within *E. coca.*

Both varieties of *E. novogranatense* are known today only as cultivated plants. Both varieties are well adapted to arid conditions and usually are grown in areas where *E. coca* could not survive. In both alkaloid and flavonoid chemistry, *E. novogranatense* differs fundamentally from *E. coca* (Bohm et al., 1982; Plowman & Rivier, 1983). Breeding experiments between *E. coca* var. *coca* and the varieties of *E. novogranatense* have demonstrated genetic differentiation among these taxa, further clarifying their specific and varietal relationships (Bohm et al., 1982).

ERYTHROXYLUM NOVOGRANATENSE VAR. TRUXILLENSE, TRUJILLO COCA

Trujillo coca is cultivated today in the river valleys of the north coast of Peru between about 200 and 1800 m elevation and in the adjacent, arid, upper Marañón river valley (Fig. 1). It is grown today on a relatively small scale for coca chewing and as a flavoring for the soft drink Coca Cola®. Although it is a highly drought-resistant shrub, it still requires some irrigation throughout its range (Plowman, 1979b).

Trujillo coca bears a leaf that is smaller, lighter green, and more brittle than leaves of *E. coca* (Fig. 4). Because it contains flavoring compounds not found in *E. coca,* Trujillo coca long has been valued for coca-flavored wines and tonics. In the last century, it was highly prized in the European and North American pharmaceutical industry for medicinal preparations (Morris, 1889; Plowman, 1982).

Today, Trujillo coca is geographically and ecologically isolated from other coca varieties, and no hybrids between them have been found. However, *E. novogranatense* var. *truxillense* has been successfully crossed with both *E. coca* var. *coca* and *E. novogranatense* var. *novogranatense.* Successful crosses were obtained in both directions between *E. novogranatense* var. *novogranatense* and *E. novogranatense* var. *truxillense.* The resulting hybrids were vigorous and vegetatively normal and exhibited morphological characters intermediate between the two parents. However, most of the hybrids between these varieties which flowered showed only 50% pollen stainability and a much reduced seed set. This suggests at least partial reproductive isolation between the varieties of *E. novogranatense* resulting from their geographical isolation in somewhat different habitats over a long period of time (Bohm et al., 1982).

Erythroxylum novogranatense var. *truxillense* also was crossed with *E. coca* var. *coca,* but with limited success. Although F_1 hybrids were obtained, these were morphologically and developmentally abnormal, and a number of them died as seedlings. They produced no flowers and clearly were ill-adapted for survival (Bohm et al., 1982). Although Trujillo coca is in several features intermediate between *E. coca* var. *coca* and *E. novogranatense* var. *novogranatense,* it is genetically much more closely related to the latter, with which it shares important chemical and ecological characters.

The leaf flavonoids of Trujillo coca reflect the intermediate nature of this variety. It shares with *E. coca* (both varieties) the 3-O-arabinosides of kaempferol and quercetin, which are absent in *E. novogranatense* var. *novogranatense.* However, both varieties of *E. novogranatense* contain the rare flavonoid ombuin-3-O-rutinoside, which is absent in *E. coca* (Bohm et al., 1982).

Based upon data currently available, Trujillo coca is placed correctly in the species *E. novogranatense* but must be recognized as a distinct variety within that species because of noted differences from the Colombian variety. Based upon genetic and geographical relationships, it is highly probable that Trujillo coca evolved directly from *E. coca* var. *coca* through intensive selection for cultivation in drier habitats and possibly for the more delicate and flavorful leaves and a

FIG. 4.   Flowering branch of Trujillo coca, *Erythroxylum novogranatense* var. *truxillense,* cultivated at Collambay, Dept. La Libertad, Peru (*Plowman 5606*).

more robust, leafy habit. Trujillo coca subsequently gave rise to the Colombian variety of *E. novogranatense* in the northern Andes under similar conditions of geographic isolation and continuing human selection pressures.

ERYTHROXYLUM NOVOGRANATENSE VAR. NOVOGRANATENSE, COLOMBIAN COCA

The fourth variety of cultivated coca is *Erythroxylum novogranatense* var. *novogranatense,* or "Colombian coca." This variety is distinguished morphologically from other varieties by its bright yellow-green foliage and lack of persistent stipules. In dried leaf specimens, identifications may be more difficult and require anatomical study (Rury, 1981). Like Trujillo coca, this variety is well adapted to dry conditions and often is cultivated in the arid, inter-Andean valleys of Colombia and along the Caribbean coast (Fig. 1). However, it is also grown in moister parts of the Colombian Andes, especially at elevations of 1000 to 1800 m.

Unlike any of the other three coca varieties, Colombian coca is quite tolerant of diverse ecological conditions, and for this reason the variety was introduced widely in horticulture in the last century and distributed to many tropical countries, both as an ornamental and as a cocaine source (Plowman, 1979a, 1982). It became an important cash crop in Java during the early part of the 20th century, introduced there by enterprising Dutch colonial planters (Reens, 1919a, 1919b).

Colombian coca is isolated geographically from other coca varieties, in contrast to the more complex distribution patterns seen in Trujillo and Huánuco cocas. This isolation is accompanied by fundamental changes in flavonoid chemistry and reproductive biology of Colombian coca. In its leaf flavonoids, Colombian coca lacks the quercetin and kaempferol arabinosides found in *E. novogranatense* var. *truxillense* and *E. coca* var. *coca,* but it contains the rutinosides, including ombuin-3-O-rutinoside, which are present in *E. novogranatense* var. *truxillense* but lacking in *E. coca* var. *coca* (Bohm et al., 1982).

As mentioned earlier, Colombian coca will not cross with *E. coca* var. *coca.* It does produce vigorous hybrids with Trujillo coca, although the resulting hybrids showed reduced fertility (Bohm et al., 1982). This suggests that *E. novogranatense* var. *novogranatense* is genetically closely related to *E. novogranatense* var. *truxillense* even though some reproductive barriers between them have developed as a result of their geographic isolation. On the other hand, *E. novogranatense* var. *novogranatense* is genetically much more distant from *E. coca* var. *coca.* In their breeding mechanisms, most erythroxylums are strongly self-incompatible, distylous species. Colombian coca is exceptional in being partially self-compatible and isolated individuals may produce abundant viable seed. Self-compatibility is considered a derived state in plants with a heterostylous breeding system, a fact that favors the view that Colombian coca is the most specialized and most recently derived variety of the cultivated cocas (Bohm et al., 1982).

Colombian coca is known only as a cultivated plant and rarely, if ever, escapes from cultivation. Today it is grown on a small scale by isolated Indian tribes of the Colombian Andes, primarily in the Sierra Nevada de Santa Marta and in the Departments of Santander, Cauca, and Huila. It is not extensively cultivated for cocaine production owing to the same difficulties in extracting the alkaloid that are encountered with Trujillo coca leaves; rather, Colombian coca is employed mostly for chewing and as a household medicine. It is commonly planted as an ornamental and medicinal plant throughout Colombia.

Discovery and early cultivation of coca

A scenario for man's first discovery and cultivation of coca in the montaña has been outlined earlier (Antonil, 1978; Plowman, 1979a; Bohm et al., 1982). The palatable, relatively tender, young leaves of *E. coca* var. *coca* must have been sampled first as a famine food by groups of nomadic hunter-gatherers who early inhabited the eastern Andes. At this time, coca existed as small, scattered populations in the montaña, similar to the distribution patterns of many wild species today. The stimulant and medicinal properties of the leaves were discovered, probably more than once, during this early period of experimentation. Once the stimulating effects of the leaves were known, they were routinely gathered from the forest for daily use. Refinements in the use of coca, including sun-drying the leaves, holding them in the mouth as a quid, and the addition of an alkaline substance, gradually developed and became customary. Numerous alkaline sources have been employed in chewing coca and with other drugs such as tobacco. In the montaña, the simplest and most readily available alkaline source is the ashes prepared from a wide variety of plants (Plowman, 1980; Rivier, 1981).

As supplies of coca in the wild became insufficient to meet the needs of a growing, coca-chewing population, coca shrubs were transplanted from the wild, nearer to habitations, so that a constant supply of leaves would be available. In this context, coca must have been one of the earliest plants cultivated in the montaña and is implicated in the earliest development of agriculture in this area. The first use and cultivation of coca certainly antedates the first appearance of any archeological evidence (such as ceramic representations of coca chewers or coca-chewing paraphernalia) by several thousand years.

Archeological evidence for coca chewing

The earliest suggestion of coca chewing is found in the Valdivia Culture on the Santa Elena Peninsula in southwestern Ecuador. Here small ceramic lime containers believed to be used in coca chewing have been found that date to Valdivia Phase 4, about 2100 B.C. (uncorrected radiocarbon dating). A tradition of small, decorated lime pots extends through the Machalilla Culture to Chorrera times (300–1000 B.C.), when it reached its maximum development. A small, ceramic figurine of the Chagras style also was discovered at Valdivia that clearly represents the prominent cheek bulge of a coca chewer. This piece is dated Late Valdivia (1500–1600 B.C.) and is the earliest known example of a long Ecuadorian tradition of figurines depicting "coqueros" (Lathrap et al., 1976). Skulls containing heavy accumulations of dental calculus, interpreted as an indication of heavy coca chewing with lime, have been found in a late Chorrera cemetery on the Santa Elena Peninsula (Klepinger et al., 1977). Based on the archeological evidence, it appears that the custom of coca chewing, and perhaps coca cultivation, was fully established in the Valdivia area by 3000 B.C.

Early evidence for coca chewing has been found also on the Peruvian coast in the Late Preceramic Period 6 (1800–2500 B.C.) in the form of artifacts employed in coca chewing and possibly of actual coca leaves, although the botanical material has not been identified taxonomically. Engel (1957) reported a bottle gourd and three *Mytilus* shells, all containing powdered lime thought to be used with coca, from a burial at Culebras (Dept. Ancash). Bray and Dollery (1983) have dated this site at around 2000 B.C. Engel (1963) also found "leaves looking like coca" along with large deposits of burnt lime at the site of Asia in the Omas Valley (Dept. Lima). Asia is radiocarbon dated at 1314 $\pm$ 100 B.C. but probably dates to about 1800 B.C. (M. Moseley, pers. comm.). Patterson (1971) excavated preserved coca leaves near Ancón (Dept. Lima) in the Gaviota phase dated between 1750 and 1900 B.C.; Cohen (1978) also reported coca from Ancón with a date of 1400–1800 B.C. Coca was one of the items (along with maize and marine shells) stockpiled in a group of storage structures at Huancayo Alto in the Chillón Valley (Dept. Lima), dating between 200 and 800 B.C. (Dillehay, 1979). Unfortunately, none of these early records of preserved "coca" leaves has been botanically identified because none of the original specimens can be located.

Archeological coca leaves from much later sites, primarily burials, on the Peruvian coast have been available for study, and these all belong to the variety *E. novogranatense* var. *truxillense,* Trujillo coca (Fig. 5). These include specimens

Fig. 5. Archeological Trujillo coca leaves from an Inca cemetery, Taruga Valley, Dept. Ica, Peru. Lowie Museum of Anthropology, accession no. 16-13426. Photograph courtesy of the Lowie Museum of Anthropology.

from Vista Alegre in the Rimac Valley (Dept. Lima, approx. 600–1000 A.D.); from the Yauca Valley (Dept. Arequipa, Late Horizon); from Monte Grande in the Rio Grande Valley (Dept. Ica); and from Chacota near Arica in northernmost Chile (Late Horizon). A number of these specimens were studied anatomically and found to correspond closely with modern Trujillo coca, although generally the archeological leaves were smaller in size (Rury & Plowman, 1984). Coca endocarps referable to Trujillo coca were reported from Vista Alegre (Towle, 1961) and more recently were excavated at Chilca (Dept. Lima, Late Intermediate Period) by Jeffrey Parsons (pers. comm.).

Later evidence for coca chewing, including lime pots, lime dippers, and ceramic coca-chewing human figurines, as well as occasional preserved leaves, has been found throughout the Peruvian coast from the early ceramic period to Inca times. Both Nazca and Moche ceramics depict numerous examples of coca chewers with cheek bulges, often carrying lime gourds and dippers (Yacovleff & Herrera, 1934; Jones, 1974; Donnan, 1978; Jerí, 1980) (Fig. 6).

Following the early appearance of coca chewing throughout the Formative in Ecuador, evidence for coca chewing in the form of lime pots and "coquero" figurines are represented in all later phases up until Inca times, especially in the provinces of Manabí, Esmeraldas, and Carchi (cf. Meggers, 1966; Drolet, 1974; Naranjo, 1974; Jones, 1974; Bray & Dollery, 1983) (Fig. 7).

FIG. 6. Miniature vessel depicting the head of a man chewing coca, Sausal, Chicama Valley, Dept. La Libertad, Peru, Moche III–IV, ca. 300–600 A.D. Peabody Museum of Archaeology and Ethnology, accession no. 46-77-30/4993.

Archeological evidence for widespread coca chewing in Colombia is well documented by a great many coca-related artifacts. During the first millennium A.D., the Quimbaya culture of the middle Cauca Valley (near the modern city of Pereira) produced numerous, beautifully crafted, gold lime pots, along with gold lime dippers. Some are furnished with gold-beaded necklaces so that they may be worn. In addition, gold figurines carrying gold lime pots in their hands have been recovered from this culture area (Jones, 1974; Antonil, 1978; Bray, 1978; Hemming, 1978). Ceramic lime pots representing coca chewers are also known from Colombia (Fig. 8).

In the San Agustín culture in the Department of Huila in southern Colombia, a number of monolithic statues have been found that strongly suggest coca chewing by the presence of extended cheek bulges and small bags (for coca leaves) slung across their chests (Pérez de Barradas, 1940; Uscátegui, 1954; Reichel-Dolmatoff, 1972; Antonil, 1978) (Fig. 9). One partially destroyed statue known as "El Coquero" at El Tablón in the valley of San Andrés de Pisimbalá distinctly shows a

FIG. 7. Tairona miniature jar depicting seated figure with large cheek bulge, probably a lime pot representing a coca chewer, San Pedro de la Sierra, Ciénaga, Dept. Magdalena, Colombia. Museo de Oro, Bogotá, accession no. CT 1383. Photograph by Robert Feldman.

small pouch hanging from one side and a lime gourd from the other (Antonil, 1978). The San Agustín statues are dated approximately to the first millennium A.D. The town of San Agustín long has been, and continues to be, a major center of coca cultivation and distribution in the upper Magdalena Valley.

FIG. 8.   Ceramic figures of coca chewers of the Capulí style from Dept. Nariño, Colombia, 800–1250 A.D. Right figure from Museo de Oro, accession no. CN 3115; left figure from Museo Arqueológico del Banco Popular, Bogotá, accession no. N-8511. Photograph by Robert Feldman.

Although there is archeological evidence that coca also reached further north into Central America, these findings are of a considerably later date than those in South America. Lothrop (1937) reported a small, carved bone head with a prominent cheek bulge from Sitio Conte in the Coclé culture of Central Panama, which is dated between 500 and 700 A.D. This figurine closely resembles figurines from Manabí Province in coastal Ecuador as well as the early Valdivia figurine discussed above. Stone (1977) mentioned small figures of gold and stone from the Diquis region of Costa Rica that show the characteristic cheek bulges of coca chewers.

Only in coastal Peru can we identify the variety of coca being employed because of the remarkable preservation of delicate plant materials in the arid desert environment. Trujillo coca appears to be present here around 1800 B.C., although it probably evolved as a distinct variety elsewhere (Plowman, 1984). We have no direct evidence from archeological leaves, but it may be presumed that *E. coca* var. *coca* was being cultivated and utilized for chewing much earlier in the east Andean montaña of Peru and Bolivia. Both Trujillo and Huánuco coca probably were used in different parts of Ecuador, where appropriate dry and wet habitats for these varieties are present. Colombian coca certainly was the variety

Fig. 9. Large triangular stone face with stylized cheek bulges suggesting coca chewing, San Agustín Culture, San Agustín, Dept. Huila, Colombia, first millennium A.D.

used in the mountains of Colombia, along the Caribbean coast and probably in Central America (Plowman, 1984).

The cultivation of coca

The several varieties of coca are grown under different ecological conditions and their methods of cultivation vary from region to region. Coca is grown on a much larger scale and in a more organized way in Peru and Bolivia than in Colombia or in Amazonia, where until recently there has been little commercial production.

COLOMBIAN COCA

Colombian coca (*E. novogranatense* var. *novogranatense*) is produced in relatively small plots, averaging perhaps one-half hectare (Antonil, 1978). Plantings are laid out on flat or gently sloping areas rather than on steep slopes as practiced in Peru and Bolivia. Although most plantations are found between 1000 and 2000 m elevation, the better quality coca (i.e., higher cocaine content and smaller leaf) is grown at the upper limits of cultivation. Colombian coca shrubs are allowed to grow much larger and bushier and are more dispersed within a planting, in contrast to Huánuco and Bolivian cocas which are kept relatively small and are planted in neat, straight rows.

Fig. 10. Plantation of Trujillo coca, *Erythroxylum novogranatense* var. *truxillense*, growing under shade of *Inga feuillei* in its arid habitat at Collambay, Dept. Trujillo, Peru.

Colombian coca is grown exclusively from seed that is gathered in conjunction with the main harvests. The seeds are planted immediately in a shallow seed bed. When the young seedlings emerge, they are shaded from direct sun; they are not planted out until they are 20 to 30 cm tall. Plants are ready for harvesting after about two years. Depending on local conditions of climate and soil, leaves may be harvested two to three times per year. Each bush produces about 500 g of leaves per harvest (Bejarano, 1945). Throughout the mountains of Colombia (in contrast to the Colombian Amazon), coca is picked exclusively by women and children. Only the mature leaves are harvested, and they must not be overripe (Antonil, 1978). The middle-aged to oldest leaves on a branch have the highest cocaine content, whereas the youngest leaves have considerably lower values (Rivier, 1981). During harvesting, a branch is held in one hand while the ripe leaves are picked off one at a time with the other hand. As in the harvesting of all varieties of coca, it is very important not to damage or break the terminal buds on the twigs, since these will furnish the flush of leaves for the subsequent harvest.

TRUJILLO COCA

In coastal Peru and the upper Marañón valley, Trujillo coca (*E. novogranatense* var. *truxillense*) is cultivated in relatively flat areas along the valley bottoms known as "playas," where the fields can be irrigated from rivers by means of irrigation canals or "aséquias." Trujillo coca shrubs resemble Colombian coca in habit and are allowed to grow relatively large and bushy, with ample space left between

Fig. 11.   Terraced plantations of Bolivian coca (*Erythroxylum coca* var. *coca*) in the Yungas region between Coroico and Ayapata, Dept. La Paz, Bolivia (*Plowman & Davis 5179*).

each plant. Fields must be irrigated regularly because Trujillo coca is grown exclusively under arid conditions; however, it is remarkably resistant to drought and survives long periods when no irrigation water is available. Because of the intense and desiccating solar radiation in the Peruvian desert, Trujillo coca often is provided with up to 50% shade by being planted under the leguminous tree *Inga feuillei* DC. (Fig. 10).

Trujillo coca is harvested similarly to Colombian coca, but because about 75% of the Trujillo crop is sold for industrial purposes (Coca Cola® production), the leaves may be harvested less carefully, with an entire branch being stripped of its leaves in one fell swoop. Trujillo coca may be harvested three times a year, usually in December, March or April, and July. During the Peruvian winter (June to September), the shrubs grow very slowly and produce the smallest crop of the year.

HUÁNUCO COCA

Huánuco or Bolivian coca (*E. coca* var. *coca*) is cultivated along the eastern flanks of the Andes from northern Peru to Bolivia (Figs. 1 & 3). This is an area of generally high rainfall and fertile soils, covered naturally by moist tropical forest. Although this variety is cultivated generally between 500 and 2000 m, the best quality and highest yields are produced at 1000 to 1500 m. The highest cocaine content in Huánuco coca is found in plants grown above 1500 m, but plants grow much slower at this altitude.

Plantations of *E. coca* var. *coca* may be constructed with or without terraces on steep mountainous slopes ("coca de la altura") (Figs. 11 & 12) or on flatter areas without terracing along valley bottoms ("coca de la playa"). Coca requires a well drained soil; plantations on slopes are preferred and produce a better quality, stronger leaf. However, in some drier valleys, such as La Convención in Cuzco, coca de la playa has the advantage of available irrigation during the more marked dry season (July to October) (Gade, 1975).

New plantations, with or without terracing, are carefully constructed on newly cleared land (Fig. 12). Because of the high rainfall and steep slopes throughout much of the eastern Andes, soil erosion is a serious problem. Terraces constructed along the contours of the slopes help to prevent excessive run-off, but these must be constantly maintained. The best constructed terraces are found in the Yungas coca-growing region of Bolivia, but even here erosion has destroyed many areas for further production and many extant "cocales" are planted on poor rocky subsoils (Fig. 13).

After a field is prepared, new plantings are started in at least two different ways. Traditionally, seeds culled from a recent harvest are planted in a protected, shaded nursery ("almáciga") with a light covering of fine soil. The young seedlings gradually are exposed to more and more sunlight, and after about three to four months, they are ready to be set out in rows in the fields. Seedlings are planted fairly close together to allow for later thinning of unhealthy or diseased plants and for natural attrition. Density of plants varies from place to place. In the Peruvian departments of Ayacucho and Cuzco, coca shrubs are allowed to grow taller and are spaced further apart than in the Bolivian Yungas, where plantations resemble rows of low coca "hedges." In Cuzco and Huánuco, the young coca seedlings in a new plantation may be interplanted with manioc (*Manihot esculenta* L.), which serves as protective shade during the first nine months or so of growth.

In Huánuco, coca seeds may be planted directly out in the fields. Shallow holes 20 cm square and 40 cm apart are dug in rows running up and down the hillsides without terracing (Fig. 14). Manioc is often pre-planted in anticipation to provide shade for the young coca seedlings. Seeds are planted directly in the holes and thinned eventually to four plants per hole, which then are allowed to grow up in place. Because there is no terracing and plants are planted in rows running up and down the slopes, erosion is especially serious in Huánuco. Topsoil from a newly prepared field soon washes away. As the roots of the shrubs become exposed with the heavy rains, soil from between the rows is heaped up around the plants, resulting in gullying of the fields (Fig. 15). Although coca will produce surprisingly well under these conditions, the lateritic soil eventually becomes hard and will support little or no vegetation, including aggressive weeds, after coca is taken out of production.

Once established, a plantation of *E. coca* var. *coca* will start producing after one to two years and reach maximum productivity in about five years. If plantations are well maintained by weeding and erosion control, they may be productive for 40 years or more, although productivity decreases after 10 to 15 years (Albo, 1978). In most areas where *E. coca* var. *coca* is grown, three to four harvests a year are possible. In most areas, there is little or no fertilization of the plantations. In ecological terms, coca is an ideal crop for the steep, wet slopes of the eastern Andes since it is able to survive and remain productive for years on heavily leached soils that will support no other crop plants. In Huánuco, owners of large coca plantations ("fundos") employ modern agricultural methods by fertilizing their plantations and applying herbicides and insecticides (Plowman & Weil, 1979). These well managed cocales, which largely supply the clandestine cocaine market, may produce up to six crops a year.

FIGS. 12–13. 12. Newly constructed terraces for planting coca (*Erythroxylum coca* var. *coca*) in the Apurímac Valley, Dept. Ayacucho, Peru. 13. Newly planted seedlings of Bolivian coca (*Erythroxylum coca* var. *coca*) in highly eroded soils in the Yungas region near Ayapata, Dept. La Paz, Bolivia.

FIG. 14.   Typical construction of a new plantation of Huánuco coca (*Erythroxylum coca* var. *coca*)
near Tingo María, Dept. Huánuco, Peru. Seeds are planted directly into the fields in shallow, square
pits excavated in vertical rows on hillsides.

$\longrightarrow$

FIGS. 15–16.   15. An old plantation of Huánuco coca (*Erythroxylum coca* var. *coca*) showing highly
leached soil and gullying of fields, Tingo María, Dept. Huánuco, Peru. 16. Sun-drying of leaves of
Trujillo coca (*Erythroxylum novogranatense* var. *truxillense*) at Collambay, Dept. La Libertad, Peru.

Fig. 17.   Mixing leaves of Trujillo coca (*Erythroxylum novogranatense* var. *truxillense*) from different areas of northern Peru at the warehouses of ENACO in Trujillo.

Production yields vary considerably from area to area. In Peru, yields in 1971 varied from 410 kg/hectare (Dept. Madre de Dios) to 1200 kg/hectare (Dept. San Martín), with a national average of 810 kg/hectare. Yields of Trujillo coca generally were higher than those of Huánuco coca (Daneri Pérez, 1974). In Bolivia, 1972 yields in the traditional coca districts of the Yungas averaged only 260 kg/hectare, whereas the relatively new Chapare districts averaged 851 kg/hectare (South, 1977).

AMAZONIAN COCA

In contrast to other varieties of cultivated coca, Amazonian coca (*E. coca* var. *ipadu*) is grown from stem cuttings. Entire plots may be derived from a single clone, and fertile seed rarely is produced. Vegetative propagation of Amazonian coca is an adaptation to the shifting slash-and-burn agriculture that is practiced among tribes in the Amazonian lowlands. Soils are poorer than in the Andes, and new fields must be cleared every two or three years. Stem cuttings up to 30 cm long and one cm in diameter are merely inserted into the ground of a newly cleared and burned field. Root formation is rapid, and within six weeks the new plants have leafed out. Plants are ready for first harvesting after about six months (Plowman, 1981).

The preparation of coca leaves for chewing

After harvesting, coca leaves of all varieties must be dried quickly and completely to preserve their flavor and texture for chewing and their alkaloid content for chewing and for cocaine extraction. Techniques of drying vary considerably depending on variety and geographical area.

COLOMBIAN COCA

Colombian coca (*E. novogranatense* var. *novogranatense*) always is dried by toasting in ceramic pans over a slow wood fire while constantly turning the leaves to prevent burning. The characteristic bright yellowish green color of the leaves changes to a yellowish brown during this process. The strong aroma of methyl salicylate present in the fresh leaves is largely lost during toasting and is replaced by a grassy, smoky flavor. When the leaves are completely dried and removed from the pan, they are extremely brittle and cannot be chewed in this state. As in the case of all coca in which the leaves are chewed whole, it is necessary to allow them to reabsorb ambient humidity until they become soft and pliable. This sometimes is referred to as "sweating." After it is picked, Colombian coca may be packed into large sacks and left to ferment slightly overnight before it is dried. This technique, along with pan-toasting, alters the taste as well as the chemical composition of the leaf, but the details of these chemical changes are unknown.

Most Colombian coca is consumed locally for the purpose of chewing by Indians and mestizos. Only in southern Colombia is there any significant commerce in coca leaves for chewing, and then only on a small scale. In spite of the problems of extracting cocaine from Colombian coca (Plowman & Rivier, 1983), there exists some illicit production in both the Sierra Nevada de Santa Marta and in the southern mountains. Most cocaine exported from Colombia, however, is processed there from crude cocaine paste manufactured in Peru, Bolivia, and recently the Amazon (see below under Huánuco coca).

TRUJILLO COCA

Trujillo coca (*E. novogranatense* var. *truxillense*), owing to the hot, dry climate where it is grown, always is sun-dried. The leaves are laid out on large cement or earthen patios and constantly turned until they are completely dry (Fig. 16). During the drying process, Trujillo coca emits an intense odor of wintergreen, which is immediately noticeable in storage rooms, even days after drying has been completed. The odor, however, diminishes rapidly and often is gone by the time the leaves reach the highland markets.

Trujillo coca represents about 6% of the "official" Peruvian crop, not counting illicit coca production for cocaine manufacture. Of all areas where coca is produced in Peru, those areas of Trujillo coca are most closely controlled by the government coca monopoly, the Empresa Nacional de la Coca (ENACO). Plantations are carefully overseen by ENACO officials, especially in the Department of La Libertad near Trujillo. Relatively little of the Trujillo crop is used for cocaine manufacture, but we have almost no information on Trujillo coca production in the remote areas of the upper Marañón valley.

Seventy-five percent of the Trujillo crop is destined for export to the United States for extraction of flavorings for the soft drink Coca Cola® and, as a byproduct, for the extraction of pharmaceutical cocaine. Trujillo coca from the entire growing area of this variety is shipped to ENACO warehouses in Trujillo, where leaves from different areas are mixed together to produce a more uniform product (Fig. 17). These are then packed into bales of 80 kg each and shipped to New York from the port of Salaverry near Trujillo (Fig. 18). In 1970, over 450 metric tons of Trujillo coca were exported.

FIG. 18. Packing leaves of Trujillo coca (*Erythroxylum novogranatense* var. *truxillense*) into bales at the ENACO warehouses in Trujillo, Peru. These leaves are destined for export to New York for use as flavorings in Coca Cola.

FIG. 19. Large drying oven (secador) for drying leaves of Huánuco coca (*Erythroxylum coca* var. *coca*) on the outskirts of Tingo María, Dept. Huánuco, Peru.

HUÁNUCO COCA

Leaves of montaña-grown coca (*E. coca* var. *coca*) may be sun-dried or oven-dried. Depending on local climate and the time of year, leaves may be laid out on open patios to dry in the sun like Trujillo coca. Because of the constant threat of rains in many parts of the montaña, the crop frequently is damaged by moisture. If leaves become wet during the drying process, they quickly begin to ferment, turning brown and becoming highly unpalatable. Leaves of *E. coca* var. *coca* are most susceptible to deterioration of any of the varieties and rapidly undergo chemical changes during fermentation in which malodorous amines are produced. The presence of high levels of aromatic oils in Colombian and Trujillo coca may retard or prevent this decomposition, since these varieties deteriorate less rapidly. Coca that has been poorly dried is considered of lowest quality and commands the lowest price; it is unsuitable for chewing and has lost most of its alkaloid content.

Because of the problems of drying large amounts of coca in the montaña, commercial growers in Huánuco now rely on large drying ovens ("secadores") fired by wood (Fig. 19). These often are two- or three-storied buildings with many layers of racks covered with porous cloth on which the leaves are placed. The fire is built in a furnace at ground level so that heat rises throughout the building. Large quantities of leaves can be dried thoroughly and quickly in as few as 12 hours. There are numerous secadores in the coca districts around Tingo María, and small growers often sell their fresh coca to the owners of the secadores for drying. A number of the larger and more conspicuous secadores around Tingo María have been closed by the authorities because of their association with illicit

Fig. 20.   Preparation of coca pisada in which leaves of *Erythroxylum coca* var. *coca* are pounded or trampled prior to drying to create a distinctive flavor, Mantaro Valley, Dept. Huancavelica, Peru. Photograph courtesy of Oscar Tovar.

cocaine production. Coca production in Huánuco far exceeds that which is required by native chewers.

High quality, montaña-grown coca is recognized by its light to medium green color and fresh, "coca" odor and flavor. Huánuco coca is prized for its uniform, intact, freshly dried leaves, which result from their being oven-dried and then sifted to remove smaller and broken leaves (Weil, 1976). There is considerable commerce in coca leaves for native chewing within Peru and Bolivia, since highland chewers often prefer leaves from one growing district or another. As with any vegetable product, coca consumers are highly sophisticated in their appreciation of coca quality and variety and most have specialized individual preferences.

One interesting variation in the preparation of montaña coca is known as "coca pisada" or "trampled coca." This process is employed in southern Peru in the coca-growing districts near Cuzco, in the Mantaro valley in Huancavelica, and probably elsewhere. Freshly harvested coca leaves are spread out on the drying patio or ground and then trampled by one or two barefoot workers or beaten with sticks or special pounders for about half an hour (Fig. 20); they are then dried normally in the sun. The pounding procedure causes the leaves to develop a dark, brownish color and special flavor, which is preferred by some chewers (Bües,

FIG. 21. Coca vendors in the central market in Cuzco, Peru. The various bags contain coca of different quality and from different areas of southern Peru and northern Bolivia.

1911; Weil, 1976). This produces a type of fermentation in the leaves, but one that differs from fermentation caused by spoilage. Coca pisada is sold in the markets of Cuzco as "Cuzco negra" along with "Cuzco verde" and other varieties (Fig. 21).

As soon as montaña-grown coca is thoroughly dried, it is allowed to "sweat" to become pliable. It then is packed into large bales weighing about 60 kg. In Huánuco, the baling material is made from a specially woven, coarse woolen fabric known as "jerga"; in Cuzco, a muslin cloth is used. The choice of material is important because the packed leaves need to be protected from the elements during transportation, but at the same time must "breathe" to prevent fermentation. In commercial growing areas such as Cuzco and Huánuco, the leaves are pressed into bales with large, mechanical, hand-driven presses.

The faster the leaves destined for chewing are transported to the dry and cold high-altitude Sierra, the better their quality and commercial value are preserved. The serious problem of deterioration of coca during transport from the field to markets long has been recognized and was one of the chief obstacles to the introduction of coca to Europe in the 19th century (Lyons, 1885; Squibb, 1885; Rusby, 1888; Morris, 1889; Mortimer, 1901). The fantastic claims by South American explorers for the virtues of coca leaves could not be matched by the stale and moldy leaves that reached Europe after a months-long sea voyage. Traditionally, coca was transported from the montaña to the highlands by llama caravans or by human bearers; this largely has been replaced today by trucks and in some cases mule trains.

Throughout the montaña, a growing percentage of coca production is diverted for making crude cocaine paste or base ("pasta"), which is readily prepared from dried leaves under primitive conditions in the areas of cultivation. Clandestine factories ("cocinas" or "pozos") are numerous in the expanding coca districts of Tingo María in Peru and Chapare and Santa Cruz in Bolivia. Only a few common chemical reagents are necessary for extracting cocaine paste, including sodium carbonate, kerosene, and sulfuric acid. Dried coca leaves are reduced in bulk by 200 to 400 times in making the paste, which is in turn easily transported to more sophisticated laboratories in urban areas, where the paste is further purified into cocaine hydrochloride. No one knows the extent of illicit coca production in the montaña. Estimates vary wildly from one source to another, and no recent estimates appear to be reliable because of a tendency of government agencies to underestimate or overestimate production, depending on their vested interests. For example, in Peru alone estimates of coca production vary between 20 and 50 million kg of leaves per year. In 1974, Bolivia produced 12 million kg which was double the production for 1971 (South, 1977). For 1978, Bolivia officially reported a production of 19.5 million kg of coca leaves, but the amount consumed for chewing was not known (United Nations, 1980).

AMAZONIAN COCA

In the western Amazon, the leaves of *E. coca* var. *ipadu,* Amazonian coca, are dried by toasting over a slow fire in special ceramic bowls or pans (Fig. 22), a method similar to that employed for Colombian coca. Amazonian coca leaves are harvested and prepared daily because of the rapid spoilage of coca in the tropical lowlands. The leaves are toasted to dryness, then reduced to a fine powder by pounding in a special mortar and pestle ("pilón"), followed by careful sifting (Prance, 1972; Plowman, 1981; Schultes, 1981) (Fig. 23). The reason for preparing Amazonian coca as a powder, in contrast to the chewing of whole coca leaves elsewhere, probably is a consequence of the larger, unwieldy leaf size and low cocaine content of Amazonian coca (Plowman, 1981; Plowman & Rivier, 1983).

Until recently, there was no commercial production of Amazonian coca, and it was virtually unknown except to a handful of botanists and anthropologists working in the Northwest Amazon. However, in the mid-1970's, Colombian cocaine traffickers discovered coca in use among certain Amazonian tribes. Although the Amazonian variety is much lower in cocaine content than Peruvian and Bolivian montaña coca, the Colombians found that it was easier to extract cocaine from the Amazonian variety than from the traditional coca grown in the mountains of Colombia (i.e., *E. novogranatense* var. *novogranatense*). Cocaine entrepreneurs moved into those areas of Amazonian Colombia where coca was used traditionally on a small scale by a few Indian tribes. Encouraged by strong economic incentives, these tribes began growing more coca and selling it to the Colombian nationals, who began making not only cocaine paste but also pure cocaine hydrochloride in Amazonian laboratories. Traditional areas of coca use in the territories of Amazonas and Vaupés were exploited first. Subsequently, Amazonian coca was taken to the "llanos" of eastern Colombia where plantations were started in remote areas of the Departments of Meta and Guaviare. The effect on the traditional cultures of the area, as well as on the traditional and healthful use of coca, has been devastating. Visitors to the area report that social and cultural disintegration has proceeded at an alarming pace, as the Colombian "mafia" has taken complete control of some indigenous areas (B. Moser, C. & S. Hugh-Jones & A. Weil, pers. comm.).

FIG. 22. Bora tribesman toasting leaves of Amazonian coca (*Erythroxylum coca* var. *ipadu*) in a special ceramic bowl, Brillo Nuevo, Río Yaguasyacu, Dept. Loreto, Peru. Note his distended cheek containing a quid of powdered coca. Photograph courtesy of Laurent Rivier.

The chemistry of coca

Although at least 15 different alkaloids have been reported from the leaves of the cultivated cocas (Willaman & Schubert, 1961; Turner et al., 1981a) and frequently are cited in the literature, their existence in the living plant recently has been questioned (Rivier, 1981; Plowman & Rivier, 1983). In a detailed study of all four varieties of cultivated coca, only cocaine and cinnamoylcocaine were measured by GC-MS (Plowman & Rivier, 1983). The natural occurrence of the other reported alkaloids in coca remains to be demonstrated with carefully evaluated methods using modern analytical techniques on fresh and/or well preserved plant materials; these other alkaloids may in fact prove to be artifacts of the storage and extraction procedures.

In the most complete alkaloid analysis of coca to date (Plowman & Rivier, 1983), the dried leaves of the four cultivated varieties were examined for alkaloids using stable isotope internal standard procedures for quantification. The leaves of *Erythroxylum coca* var. *coca* showed a mean of 0.63% cocaine (30 samples), which compares favorably with earlier reports of the alkaloid content of this

FIG. 23. Bora tribesman sifting a mixture of pulverized coca leaves (*Erythroxylum coca* var. *ipadu*) and leaf ashes of *Cecropia sciadophylla* Mart. through a cloth bag, Brillo Nuevo, Río Yaguasyacu, Dept. Loreto, Peru. Note both his cheeks are full of powdered coca. Photograph courtesy of Laurent Rivier.

variety (Ciuffardi, 1949; Machado, 1972; Holmstedt et al., 1977; Turner et al., 1981b). The sample of this variety with the highest amount of cocaine (0.96%) came from Chinchao (Huánuco, Peru), an area where coca is grown near the upper altitudinal limits of cultivation (1600–1800 m). This lends credence to the belief that, although coca grows slowly at such altitudes, it produces a more potent leaf.

Leaves of *E. coca* var. *ipadu,* Amazonian coca, contained the lowest amounts of cocaine with a mean of only 0.25% (6 samples). The cocaine content of this lowland variety is consistently low, even when grown under controlled laboratory conditions, and appears to be genetically controlled (Plowman, 1981; Plowman & Rivier, 1983).

Leaves of both varieties of *E. novogranatense* produced higher concentrations of cocaine than the "classical" variety, *E. coca* var. *coca.* Colombian coca (*E. novogranatense* var. *novogranatense*) yielded a mean of 0.77% cocaine (3 samples), and Trujillo coca (*E. novogranatense* var. *truxillense*) showed a mean of 0.72% cocaine (14 samples). The highest cocaine concentration (1.02%) of all the cultivated cocas was found in a sample of Trujillo coca (*Plowman 5600*) collected at Simbal near Trujillo, Peru. This finding contradicts an earlier belief that Trujillo coca is lower in cocaine content than other varieties (Mortimer, 1901; Machado, 1980).

Both cis- and trans-cinnamoylcocaine are found in all four varieties of cultivated coca. Cinnamoylcocaine always is found together with cocaine and never alone. Both varieties of *E. novogranatense* contained much higher concentrations of cinnamoylcocaine than either variety of *E. coca.* In both varieties of *E. novogranatense,* the amount of cinnamoylcocaines may exceed that of cocaine, although the ratios between the alkaloids varied widely from sample to sample.

Several earlier workers recognized the high percentage of what was then called "uncrystallizable cocaine," especially in Java coca (*E. novogranatense* var. *novogranatense*) (Morris, 1889; Hesse, 1891; Mortimer, 1901; Reens, 1919a). The uncrystallizable fraction of these varieties now is thought to contain the cinnamoylcocaines. Using methods employed at the turn of the century, chemists found difficulty in extracting and purifying pharmaceutical cocaine from leaves of *E. novogranatense.* Cocaine was produced from Java coca by first hydrolyzing the relatively high amounts of total alkaloid to ecgonine and then semi-synthesizing cocaine from this base.

In addition to alkaloids, coca leaves contain a wide variety of other constituents, many of which are incompletely known (Hegnauer, 1981). Both varieties of *E. novogranatense* contain high concentrations of methyl salicylate (wintergreen oil) (Romburgh, 1894, 1895; Reens, 1919a) and probably other aromatic oils that give a distinctive flavor to the dried leaves and provide the basis for the use of Trujillo coca as a flavoring in beverages. Only minute amounts of methyl salicylate have been reported from *E. coca* (Romburgh, 1894, 1895). An array of flavonoids derived from quercetin and kaempferol also have been identified in the cultivated cocas and are useful taxonomic markers. The flavonoids of both varieties of *E. coca* are identical, but those of the two varieties of *E. novogranatense* differ both from one another and from *E. coca.* Both varieties of *E. novogranatense* contain the rare flavonoid ombuin-3-O-rutinoside, which is absent from *E. coca* (Bohm et al., 1981). None of the flavonoids of coca is known to be pharmacologically active.

During the 1940's, a small group of public health officials in Peru campaigned vehemently against the native use of coca, which they perceived to be detrimental to the health of the Indians. One of their arguments was that coca chewing resulted in malnutrition because they believed that coqueros chewed coca in lieu of food (cf. Saenz, 1941; Gutiérrez-Noriega & Zapata Ortíz, 1948; Kuczinski-Godard & Paz Soldán, 1948; Zapata Ortíz, 1970). These arguments have been refuted repeatedly as unscientific (Burchard, 1975; Grinspoon & Bakalar, 1976; Carter et al., 1980a). During the 1970's, a number of studies demonstrated that coca leaves in fact contain impressive amounts of vitamins and minerals (Machado, 1972;

Duke et al., 1975; Carter et al., 1980a). In one study (Duke et al., 1975), the amounts of 15 nutrients in coca leaves were compared to averages of these nutrients present in 50 Latin American foods. Coca was found to be higher in calories (305 per 100 g vs. 279), protein (18.9 g vs. 11.4 g), carbohydrate (46.2 g vs. 37.1 g), fiber (14.4 g vs. 3.2 g), calcium (1540 mg vs. 99 mg), phosphorus (911 mg vs. 279 mg), iron (45.8 mg vs. 3.6 mg), vitamin A (11,000 IU vs. 135 IU), and riboflavin (1.91 mg vs. 0.18 mg). Based on these data, 100 g of Bolivian coca leaves would more than satisfy the Recommended Dietary Allowance for reference man and woman in calcium, iron, phosphorus, vitamin A, and riboflavin. Picón-Reátegui (1976) pointed out that vitamin A intake in Andean populations is very low, so the extremely high vitamin A content in coca leaves would supplement this deficiency significantly. However, since the time when the nutritional value of coca was proved, no researchers have conducted studies on the actual or potential contribution of coca in native diets. A quid of finely powdered Amazonian coca gradually dissolves with saliva and may be completely ingested, and the intake of nutrients in this case would be higher than in chewing whole leaves.

Coca chewing

THE MECHANICS OF COCA CHEWING

Coca leaves are chewed in a relatively uniform manner throughout their area of use, although there exist numerous minor variations. The greatest divergence from the normal pattern is found in the Amazon, where coca is used in powdered form. In the Andes, the act of chewing coca is accompanied by a complex series of rituals that are deeply embedded in traditional Quechua life. These are discussed in detail later.

Coca always is dried before use; this facilitates the rapid release of the chemical constituents from the leaves during chewing. The dried leaves are placed in the mouth one or a few at a time and slowly moistened with saliva. Almost immediately, a rich green juice issues from the leaves and they become soft and pliable. They are then moved about the mouth with the tongue and rolled into a ball or quid and pushed into one cheek. Coca is never actually chewed, but rather the moistened quid of leaves is sucked upon to extract the juices, which slowly trickle into the stomach. In South America, a number of words are used specifically to denote coca chewing: "mambear" (Colombia); "chacchar," "acullicar," "pijchear" (Peru, Bolivia); "coquear" and "mascar" (general).

The juice that emanates from the quid is distinctive in flavor and depends somewhat on the variety of coca. Generally, coca has a grassy or hay-like taste, with a hint of wintergreen in Trujillo coca. During the earliest stages of chewing, all coca varieties are distinctly bitter because of the presence of alkaloids. This bitterness is counteracted by the addition of an alkali substance, such as powdered lime or ashes—or even baking soda (sodium bicarbonate) among non-native chewers. The alkali not only "sweetens" the chew but also noticeably potentiates its effects, both in numbing the cheeks and tongue (through the anesthetic effect of cocaine) and by increasing the stimulating effect. Additional doses of alkali periodically are added to the quid to maintain its effect on the chew; more leaves also may be added until the quid reaches an optimal size for the chewer.

The amount of time the coca quid is kept in the mouth varies, depending on the individual user, from about 30 to 90 minutes, after which the quid is spat out. Amount and duration of chewing depends in part upon the cost and availability of leaves in a particular region. The amount of coca chewed also varies according to individual taste, ranging generally from 25 to 75 g of leaves per day.

Hanna (1976) suggested a mean daily consumption of 50 grams, which is comparable to a figure of 43 grams found by Ciuffardi (1949).

Few researchers have studied the pharmacology of coca chewing in vivo among native chewers. Ciuffardi (1948a, 1948b) found that upon chewing 30 grams of coca leaves, 87% of the alkaloids, with an average of 112 mg of cocaine, were absorbed. In another experiment, Ciuffardi (1949) studied the effects of the addition of "llipta," an alkali admixture made from ashes. He found that the addition of llipta to the quid significantly increased the absorption of cocaine into the bloodstream; to a lesser extent, llipta hastened the extraction of the alkaloids from the leaf material. Rivier (1981) interprets the role of llipta as "providing an alkaline medium in which the cocaine conjugates in leaf material are liberated as the free base, which is absorbed into the bloodstream more readily than the hydrochloride through the lipophilic mucous membrane of the mouth before cocaine is hydrolyzed to benzoylecgonine and ecgonine."

Because of its extreme fineness, using Amazonian coca powder is somewhat more complicated than chewing whole leaves. A heaping tablespoonful is placed in one cheek cavity while the head is bent to one side in order that the powder not enter the throat. The chewer then closes his mouth for about five minutes while he works the powder with his tongue and moistens it with saliva into a thick paste. This then is packed between the cheek and the gums with the tongue. The pasty quid gradually dissolves with saliva and is swallowed little by little until it imperceptibly melts away into the stomach (Plowman, 1981).

In the Northwest Amazon of Colombia, coca powder also occasionally may be taken in the form of a snuff (Wavrin, 1937; Schultes & Holmstedt, 1968), but the details of this method of use are unknown.

ADMIXTURES TO COCA

Coca is rarely, if ever, chewed alone. Some form of alkaline substance always is added to the coca quid, and a number of other flavoring or potentiating admixtures also may be employed. The custom of adding alkali to coca is similar to the use of powdered lime with betel nut (*Areca catechu* L.) in India and Southeast Asia or of ashes with pituri (*Duboisia hopwoodii* F. v. Muell.) in Australia and with tobacco (*Nicotiana tabacum* L.) in parts of South America (Miner, 1939). The use of alkaline substances with coca is presumed to be very ancient; the earliest probable evidence for coca chewing consists not of coca leaves, but of small receptacles containing powdered lime (Figs. 7, 24).

Alkaline admixtures to coca are of two basic types: plant ashes and lime made from burnt limestone, seashells or bone (Fig. 25). The geographical distribution of the use of these two types of alkali is of interest but is not fully understood. In the northern half of the area of coca chewing in the Andes, lime is the preferred source of alkali; in the southern half and in the Amazon Basin, plant ashes predominate. In coastal areas, such as the Caribbean coast of South America and the Pacific coast of Peru, the preferred (and expected) alkali source is marine shells, which are readily available. In the interior of Colombia and in parts of the Andes of northern Peru, limestone rocks are gathered from areas of known deposits and baked to make a crude powdered lime (Fig. 25). This lime is called "mambe" in Colombia and "ishku" in Peru. Whenever lime is the alkaline admixture used with coca, it is carried in a small- to medium-sized bottle gourd (*Lagenaria siceraria* [Mol.] Standley) with a narrow neck (Fig. 24). The lime gourd is called "poporo" in Colombia and "poro," "puru," or "ishkupuru" in Peru. In many areas, a small spatula or stick is carried with the lime gourd and is used to

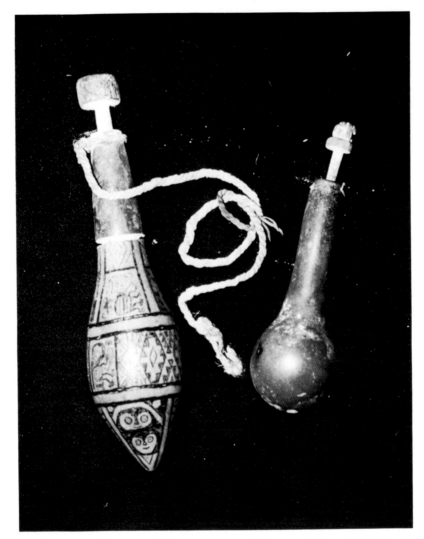

Fɪɢ. 24. Archeological lime gourds (*Lagenaria siceraria*) from coastal Peru. Left: Lime gourd with pyroengraved anthropomorphic and geometric decoration, from Hacienda Humaya, Huacho, Huaura Valley, Peru, ca. 1000–1475 A.D. Peabody Museum of Archaeology and Ethnology, accession no. 46-77-30/6189. Right: Small undecorated lime gourd with carved figure on dipstick, Cajamarquilla, Rimac Valley, Dept. Lima, Peru, date uncertain, possibly Middle Horizon, ca. 600–1000 A.D. Peabody Museum of Archaeology and Ethnology, accession no. 46-77-30/6088.

transfer the lime from the gourd to the quid in the cheek (Fig. 26). In southern Colombia, the Paez Indians merely pour the powdered lime from their gourd onto the palm or back of the hand and toss it onto the quid in the mouth. But in southern Cauca, the lime normally is not pulverized, but is used in the form of a hard lump. Small pieces are bitten off and inserted into the quid (Antonil, 1978).

In the Sierra Nevada de Santa Marta on the Caribbean coast of Colombia, Indians of the Kogi and Ika tribes continue to use coca in centuries-old, traditional patterns (Reichel-Dolmatoff, 1953; Ochiai, 1978). Only the men of these tribes

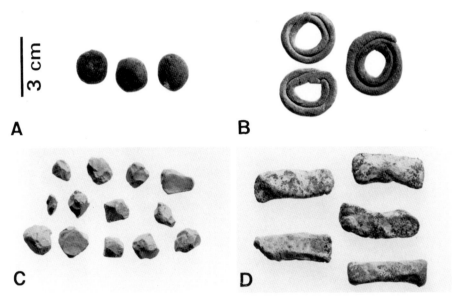

FIG. 25. Alkaline admixtures employed with coca leaves. A. Small balls of ash made from quinua stems (*Chenopodium quinoa* Willd.), Cuzco, Peru (*Plowman 7940*). B. Spirals of ash made from quinua stems, Guaqui, Dept. La Paz, Bolivia (*Plowman 7941*). C. Crude lime known as mambe, prepared by baking natural limestone, San Agustín, Dept. Huila, Colombia (*Plowman 7943*). D. "Fingers" of ash made from quinua stems and flavored with anise and raw sugar, La Parada market, Lima, Peru (*Plowman 7602*). All four samples correspond to the scale in "A."

chew coca, which they carry in elaborate woven bags slung around the shoulders. The gourds used in this area are particularly large and phallic in shape, a reflection of the sexual symbolism that coca chewing reflects among these groups. While chewing coca, the coca-laden saliva mixed with lime habitually is rubbed with the limestick around the end of the gourd. The lime precipitates as calcium carbonate and gradually builds up to form a thick rim in the form of a cylinder or disc (Mariani, 1890; Reichel-Dolmatoff, 1950; Moser & Taylor, 1965; Billip, 1979). The rim is carefully trimmed and molded and is a symbol of pride and status, since it demonstrates a man's dedication to coca chewing. A painted Mochica vessel dated about 500 A.D. from the north coast of Peru shows three coca chewers using nearly identical lime gourds with large, disc-like rims (Kutscher, 1955: 8–9). According to Jones (1974), these coqueros were thought to be "foreigners" because of their triangular cheek markings and dangling ear ornaments; whether or not they might have come from what is today Colombia is not known. However, it does suggest that the curious custom of fashioning elaborate rims on lime gourds was more widespread in the past and that there may have been cultural contacts between Peru and Colombia at an early date.

In the Andean highlands and montaña of southern Peru and Bolivia and in the Amazon basin, the preferred alkaline substance for coca chewing is made from ashes from a variety of plants and plant parts. This admixture usually takes the form of a moist black paste or, when dried, a grey rock-like substance. It is known as "llipta" or "tocra" in Peru and "lejía" or "llujta" ("llucta") in Bolivia. The

FIG. 26.   Peruvian mestizo adding powdered lime to his coca quid with a small stick. The lime is carried in the small gourd in his hand. Balsas, Río Marañón, Dept. Amazonas, Peru.

dividing line in the Andes between the northern lime-gourd-using coca chewers and the southern llipta users lies approximately at the border of the Peruvian departments of Huánuco and Pasco. Along the Pacific coast, the use of lime and lime gourds appears to be universal.

Llipta is prepared from a large number of plant species (Fig. 25). In a given area and habitat, one or a few llipta sources will be preferred. In the high Andes, the preferred plant ash comes from two species of cultivated chenopods, which are *Chenopodium quinoa* Willd., "quinua," and *C. pallidicaule* Aellen, "cañihua." Also in the highlands, llipta may be made from the roots of faba beans (*Vicia faba* L.) and from the stems and fruits of several species of columnar cacti in the genera *Cereus, Trichocereus,* and *Cleistocactus.* In the tropical montaña of the eastern Andes, maize cobs, *Musa* roots, and cacao pods commonly are burned to make llipta. In both areas numerous wild plants also are exploited and preferred locally by coca chewers. To prepare llipta, the plant is burned thoroughly to a fine ash and then mixed with water. Starchy potato water may be used to hold the ash together. The resulting pasty mass then is molded into cakes in a variety of shapes and sizes depending on local custom. Llipta may be flavored with various spices such as anise or chili peppers (Mortimer, 1901; Antonil, 1978). In order to use llipta, a small piece of the hardened cake is broken off and inserted with the fingers into the quid. One must be careful not to let the llipta touch the inner surfaces of the cheek since it may cause painful burns. The quality of llipta varies appreciably and may be extremely alkaline and caustic, or mild, very hard and rock-like, or soft and crumbly. Hard llipta dissolves slowly, and one piece may serve to supply an entire chew with alkali, without the frequent reapplications that are necessary when chewing coca with powdered lime.

In the Amazon basin and Andean foothills, ashes of *Cecropia* or *Pourouma* trees are used as the alkaline source for coca chewing. The Mashco of the southern Peruvian montaña burn the trunks of a *Cecropia* species to ashes; these are finely pulverized and stored in a bamboo tube. The ash then is added to the quid of whole coca leaves with a small stick, not unlike the use of a lime gourd and lime stick in coastal Peru (Califano & Fernández Distel, 1978). In the Beni area of northern Bolivia, ashes are prepared from the spathe of the "motacú" palm (*Scheelea princeps* [Mart.] Karst.) and are stored in a cow's horn. A small-leaved form of *E. coca* var. *coca* is chewed in this area as a quid of whole leaves to which the motacú ash is added (Le Cointe, 1934; Davis, 1983). A number of tribes in the southern montaña, including Campa, Machiguenga, Mashco, and Chimane, chew whole coca leaves with ashes. Further north in the "selva" areas of lowland Amazonia, coca always is chewed as powder, pre-mixed with finely sifted *Cecropia* or *Pourouma* leaf ashes (Plowman, 1981; Schultes, 1981) or with banana leaf ashes (Prance, 1972). Only Amazonian coca, *E. coca* var. *ipadu,* is prepared in powdered form.

Rivier (1981) measured the pH and buffer capacity of 17 different samples of coca alkali admixtures, including lime, llipta, and *Cecropia* leaf ashes, among others. The pH of these substances ranged from 10.1 in llipta made from quinua stems to 12.8 in lime made from marine shells. Llipta contains high amounts of calcium, magnesium, and potassium salts, the proportions varying according to the source (Gosse, 1861; Cruz Sánchez & Guillén, 1948; Baker & Mazess, 1963). Baker and Mazess (1963) believe that the calcium contained in llipta ingested during coca chewing is an important source of this element in the diet of coca chewers.

Besides the addition of alkaline substances, a number of other plants may be used along with coca. The most important of these is tobacco, a drug that is found almost universally among tribes that use coca. Contemporary Andean coca chewers frequently smoke cigarettes while chewing coca or even smoke as a substitute for coca when chewing is not possible (Fine, 1960). A soft tobacco paste is made in a number of areas and added to the coca quid by means of a small needle.

This custom is especially conspicuous in the Sierra Nevada de Santa Marta and in the northwest Amazon, where the tobacco paste is called "ambira" and "ampiri," respectively (Uscátegui, 1954). Tobacco paste is prepared by slowly cooking tender tobacco leaves with water; a "bush salt" made from the ashes of one of a number of plants then is added to the resulting syrupy paste. In the Amazon, the addition of tobacco paste at the beginning of a chew of powdered coca stimulates salivation and greatly facilitates the formation of a quid from the powder (Plowman, 1981). Tobacco in snuff form is used with coca by the Mashco in the Peruvian montaña (Califano & Fernández Distel, 1978) and by several tribes in Colombia (Uscátegui, 1961).

A little known but interesting coca admixture comes from a bignoniaceous vine, *Mussatia hyacinthina* (Standl.) Sandw., known as "chamairo" (Plowman, 1980; Davis, 1983). The bark of the stem of this liana is added to the quid of whole coca leaves among the Campa and Machiguenga of eastern Peru and also among the Chimane and other groups of northern lowland Bolivia. Chamairo is used as a flavoring and sweetener for the coca quid and also may be chewed alone (with ashes but without coca) as a stimulant and medicine. In northern Peru, Trujillo coca quids may be flavored with the dried leaflets of *Abrus precatorius* L., known locally as "mishquina" or "miski miski." The foliage of *Tagetes pusilla* HBK. is used with quids of *E. coca* var. *coca* in southern Peru, and the aromatic resin of *Protium heptaphyllum* (Aubl.) March is employed in the Colombian Amazon to flavor Amazonian coca powder (Schultes, 1957).

THE EFFECTS OF COCA CHEWING

The primary effect of chewing coca is a mild stimulation of the central nervous system resulting from the assimilation of cocaine from the leaves (Holmstedt et al., 1979). Some workers (Montesinos, 1965; Burchard, 1975) have suggested that the ecgonine derivatives of cocaine may play a role in the combined effects of coca chewing, but their interesting theories have not been confirmed by controlled experiments. In addition, the minor alkaloids presumed to be present in the coca leaf have been implicated in the effects of coca chewing (Mortimer, 1901; Martin, 1970), but little is known of the biological activity of these compounds. Rivier (1981) has shown that the only other alkaloid present in coca leaves at significant levels (greater than 1% of amount of cocaine) is cinnamoylcocaine, and this compound is not known to be pharmacologically active. If other alkaloids are indeed present, they exist only as trace constituents.

During coca chewing, free cocaine base is absorbed rapidly through the buccal mucosa in the mouth and to some extent in the gastrointestinal tract. Cocaine is measurable in blood plasma five minutes after coca chewing begins, which gives a measure of the rapidity of cocaine assimilation. Peak levels in plasma are reached one to two hours after chewing begins (Holmstedt et al., 1979), although the major subjective effects are felt within the first half hour of chewing. Peak blood levels of cocaine ingested during coca chewing are highly variable and depend upon several factors, including dose and concentration of cocaine in the leaf material, absence or presence of alkali admixtures, and individual experience of the chewer, among others. Blood levels of cocaine during chewing may approximate, but generally are lower than, those found after intranasal administration of cocaine (cf. Javaid et al., 1978; Holmstedt et al., 1979; Paly et al., 1980). Surprisingly, no modern detailed pharmacological studies of coca chewing in native coca chewers yet have been conducted, although numerous such studies have been conducted on cocaine users.

The stimulation experienced during coca chewing gives a sense of increased energy and strength, a suppression of the sensation of fatigue, an elevation of mood or mild euphoria, and a sense of well being and contentment. Coca also produces a temporary loss of appetite. Owing to the release of cocaine in the mouth during chewing, there is a pronounced numbing sensation of the cheeks and tongue, which results from the anesthetic action of cocaine. There is no evidence that coca chewing results in tolerance or physiological dependence, nor does it show any acute or chronic deleterious effects (Weil, 1975; Grinspoon & Bakalar, 1976; Carter et al., 1980a).

Even though cocaine is the principal and most powerful constituent of coca leaves, the complex effects of chewing coca leaves, especially those that are exploited in medicine, cannot be equated with the comparatively straightforward effects of using cocaine. As mentioned earlier, coca is a complex mixture of chemicals, including alkaloids, essential oils, flavonoids, vitamins and minerals, and other natural leaf constituents, many of which still never have been examined in coca. For example, coca has a soothing effect on disorders of the stomach and intestinal tract and is used in folk medicine for a wide spectrum of complaints. Montesinos (1965) suggested that ecgonine, a breakdown product of cocaine, may relax directly intestinal smooth muscle, and the beneficial effects on digestion of the volatile oils, such as methyl salicylate, are well known. Furthermore, coca stimulates oral secretions and may change secretion in other parts of the gastrointestinal tract (Weil, 1981). In spite of these possibilities, coca's mechanism of action on the gastrointestinal tract remains unknown.

Burchard (1975) and Bolton (1976) have suggested that coca chewing affects carbohydrate metabolism among Andean coca chewers, who typically live on high starch diets. Burchard believes that coca may protect against the development of hyperglycemia and of reactive hypoglycemia following oral glucose loads ingested by Andean chewers and suggests that this effect may involve ecgonine, one of the products of cocaine hydrolysis. Although experimental evidence for these metabolic effects is lacking, Weil (1981) suggests that coca be tested as a possible treatment for diabetes.

As many workers have pointed out, it is completely erroneous to equate the pharmacological effect of coca chewing with that of the use of highly concentrated cocaine (Mortimer, 1901; Martin, 1970; Grinspoon & Bakalar, 1976; Weil, 1981). However, until the complex chemistry of coca leaves and the pharmacology of their constituents are studied in detail, the highly beneficial, yet subtle, medicinal and restorative effects of coca remain unsubstantiated by modern medical studies.

Uses of coca as a stimulant and medicine

Whether in the high Andean altiplano or in the Amazonian lowlands, coca is employed principally for work (Burchard, 1975; Carter et al., 1980a; Plowman, 1981). Workers will take several breaks during the daily work schedule to rest and chew coca, not unlike the "coffee break" in Western society. Coca chewers maintain that coca gives them more vigor and strength and assuages feelings of hunger, thirst, cold (in the highlands), and fatigue. Coca is chewed by rural people in all kinds of professions that require physical work, especially farmers, herders, and miners in the highlands and by farmers, fisherman, and hunters in the lowlands. Coca is especially highly regarded when making long journeys on foot, both through the rugged mountains of the high Andes and through the Amazonian forests. It rarely is possible to carry adequate supplies of food on such treks, and coca is considered the best form of sustenance; this fact was recognized by the

earliest European observers in South America (cf. Mortimer, 1901; Martin, 1970).
In such situations, coca temporarily postpones the necessity for food, but it never
takes the place of food. Even today, coca is preferred by long-distance truck drivers
in the Andes to keep them alert on dangerous mountain roads and to sustain
them for long periods.

Miners in Peru and Bolivia always have depended on coca to protect them
during their unhealthy and exhausting work. After an initial period of condem-
nation and prohibition of coca, the early Spanish administrators realized that
only with coca could the Indians be forced to work in the silver mines. Miners
believe that coca helps them in a number of ways: as an energizer, as a filter
against the penetrating dust and toxic gases, as a stimulant to combat drowsiness,
and as an almost magical substance that reduces hunger (Carter et al., 1980a,
1981). Under the harsh environmental conditions in the high Andes and lowland
Amazon, coca chewers believe that only coca gives them the strength to do their
work, to maintain good health, and to protect them from disease.

The second most important use of coca is as a medicine, and this use is inex-
tricable from the Indians' belief that coca is a protector and preserver of health.
It is significant that many South Americans, Indians and non-Indians, who do
not regularly chew coca leaves as a stimulant, will cultivate the plant and use the
leaves medicinally. As an internal medicine, coca is both taken as an infusion
and chewed as a quid. Probably the most important medicinal use of coca is for
problems of the gastrointestinal tract. It is the remedy of choice for dysentery,
stomachache, stomach ulcers, indigestion, cramps, diarrhea, and other painful
conditions (Martin, 1970; Fabrega & Manning, 1972; Hulshof, 1978; Carter et
al., 1980a, 1981; Weil, 1981; Grinspoon & Bakalar, 1981). Coca also is used
commonly, by Indians, mestizos and foreigners alike, for the treatment of the
symptoms of altitude sickness, or "soroche," which include nausea, dizziness,
cramps, and severe headaches. A related use of coca is to counteract motion
sickness, a use that has received little attention in the literature (Weil, 1981).
Owing to the anesthetic effects of cocaine, coca leaves are an excellent home
remedy for toothache (Hulshof, 1978). Coca also serves as a dentifrice, and it is
commonly believed that coca helps to protect teeth and gums from decay and
disease and to keep the teeth white (Martin, 1970; Weil, 1981). Coca frequently
is used to ease rheumatic pains, taken both in an infusion and simultaneously as
a poultice over the affected part (Martin, 1970; Hulshof, 1978; Carter et al., 1981).
Coca poultices also are applied externally for headaches, sore throats, wounds,
broken bones, and irritations to the eyes. Coca also is widely employed for numerous minor and miscellaneous ailments, such as hangovers, hemorrhage, amen-
orrhea, asthma, constipation, and general debilitation (Gagliano, 1979; Grinspoon
& Bakalar, 1981; Weil, 1981). Of special importance to the Indian, coca is an
extremely valuable remedy for a number of Andean "folk" or "traditional" illnesses, which lie outside the realm of Western medicine yet play a major role in
the Andean medical belief system (Fabrega & Manning, 1972; Carter et al., 1980a,
1981). In Peru, these illnesses include "soka," a condition of weakness, fatigue,
and malaise; "fiero," a chronic wasting disease; "locura," severe mental disturbances; and others. Similar illnesses, often attributed to supernatural or magical
causes, are recognized in Bolivia. Coca, often used in conjunction with other
medicinal herbs, is a primary remedy in treating such disorders (Carter et al.,
1980a, 1981). The importance of coca in relation to these diseases is closely
associated with its reputed magical properties and role in religious life.

Since the turn of the century, the importance of coca as a medicinal plant largely
has been ignored by Western scientists, who identified the coca leaf with cocaine

and preferred to experiment with the pure, isolated compound. As a result, coca leaves completely disappeared as a pharmaceutical product and no longer were available for investigation in the United States or in Europe. Ironically, even today physicians' narcotics licenses in the United States clearly state that they have permission to dispense coca leaves. In the mid-1970's, interest in the therapeutic value of coca was rekindled among scientists as part of a general reawakening of interest in coca. Today coca again is being studied for possible applications in modern medicine. Weil (1981) has recommended that coca be studied for several therapeutic applications, including: 1, for painful and spasmodic conditions of the entire gastrointestinal tract; 2, as a substitute stimulant for coffee in persons who suffer gastrointestinal problems from its use or who are overly dependent on caffeine; 3, as a fast-acting antidepressant and mood elevator without toxic side effects; 4, as a treatment for acute motion sickness; 5, as an adjunctive therapy in programs of weight reduction and physical fitness; 6, as a symptomatic treatment of toothache and sores in the mouth; 7, as a substitute stimulant to wean users of amphetamines and cocaine from those drugs, which are more dangerous and have higher abuse potential; and 8, as a tonic and normalizer of body functions.

The role of coca in religious and social life

Coca plays a central role in the daily lives of many different groups of South American Indians, not only as a stimulant and medicine, but also as a unifying cultural and religious symbol. The very act of chewing coca in Andean communities is an ancient and basic cultural tradition by which the Indian identifies and reaffirms his or her place in the world. It should be noted that in many areas, such as the Sierra Nevada de Santa Marta in Colombia and in the Amazon basin, women are forbidden by custom to chew coca, but in other areas, especially in the high Andes of Peru and Bolivia, women use coca with as much relish as men.

In Peru and Bolivia, the traditional act of chewing coca involves a complex series of personal rituals and etiquette. The first step is to select two or three leaves from the coca bag. These are known as "k'intu." They are carefully placed one on top of the other between the thumb and index finger. The k'intu is brought in front of the mouth and blown lightly upon, and simultaneously the coquero invokes the local gods and spirits of the hills and sacred places around him. This act is known as "pukuy." The leaves then may be used to form a quid for chewing or may be crushed and blown away with additional prayers and incantations (Gifford & Hoggarth, 1976; Wagner, 1978). Wagner (1978) has described how these seemingly simple ritual acts of using coca serve to orient the Quechua Indian in a broader cultural context of time and space and in his religious studies and social affairs.

In traditional Andean communities, coca is present at nearly every public and private event or activity. It is a requisite symbol of friendship and good faith at all popular and religious festivals, engagements and weddings, baptisms, funerals, inaugurations of public officials, and formal and informal meetings at which contracts are formalized and business arrangements made (Quijada Jara, 1950; Frisancho Pineda, 1973; Gifford & Hoggarth, 1976; Carter et al., 1980a). Offerings of coca are necessary to propitiate the gods on many occasions, such as the planting of crops, insuring a productive harvest, or laying the cornerstone for a new house (Martin, 1970). There is essentially no domestic or social act that is not solemnized by making offerings of, or by chewing, coca (Quijada Jara, 1950). Coca is considered a spiritual protector for traveling in unfamiliar territory where strange and malevolent spirits abound (Quijada Jara, 1950; Wagner, 1978).

Coca always has been a major means of exchange in trade networks throughout the Andes, particularly between the tropical montaña and the high sierra and altiplano regions. Such trade networks apparently are descended from Inca times or earlier (Burchard, 1974). Long-distance trade in coca became even more extensive during the Colonial period after the Spanish took control of coca production to supply the silver mines at Potosí in Bolivia (Gagliano, 1960).

Coca is a medium of exchange not only of products but also as a symbol of friendship. Wherever coca is chewed, exchanges of coca leaves or coca powder are considered the most gracious form of greeting when people meet while traveling. Such exchanges form an immediate bond of friendship and trust and are accompanied always by the usual formalities of coca etiquette. Gifts of coca are often offered by a young man to a girl's parents to obtain their consent for marriage (Martin, 1970), and bundles of coca will be included in the dowry (Gifford & Hoggarth, 1976).

The religious and shamanistic use of coca probably is very ancient and originated from the psychoactivity produced by chewing the leaves (Martin, 1970). Ritual coca chewing enabled shamans and priests to meditate, to enter trance-like states, or to communicate with the supernatural world, even though coca produces slight mental distortion compared to hallucinogenic plants such as *Datura* and *Banisteriopsis* or even tobacco.

Many sacred practices associated with coca chewing have disappeared among tribes whose numbers were decimated or who lost their cultural identity after the Spanish Conquest. However, such ceremonies involving coca still exist among the Kogi and related tribes of the Sierra Nevada de Santa Marta in northern Colombia. These rituals have been carefully documented by the Colombian anthropologist Reichel-Dolmatoff (1950). Only the men among the Kogi are permitted to cultivate and chew coca although women are responsible for harvesting the leaves. Kogi men describe the most important effect of coca chewing as mental lucidity, which they value for ceremonial meetings, personal rituals, and religious activities in general. They assert that coca makes their bodies tingle and refreshes their memory so that they can speak, chant, and recite for hours on end. They consider the suppression of hunger caused by coca chewing a great advantage but not because they lack food. To the Kogi, fasting is a prerequisite for all religious ceremonies, and by consuming only coca, they are better able "to speak of the Ancients." According to Reichel-Dolmatoff, "the ideal of the Kogi male would be to eat nothing but coca, to abstain totally from sex, never to sleep and to speak all his life of the 'Ancients', that is, to chant, to dance and to recite."

In both Andean and Amazonian cultures, reverence for coca is reflected in its widespread use in divination, both for shamanistic healing practices and for predicting the future. These two general applications of divination are inextricably linked together in daily life. Divination is a very ancient custom among South American Indians and, in spite of relentless persecution by the Spanish clergy following the Conquest, it remains widely practiced today. The Andean Indian relies on divination for many purposes but primarily for diagnosing disease and finding a cure, for predicting the outcome of economic situations and future events in general, and for assuaging his constant fears of the spirit world which surrounds him (Contreras Hernández, 1972; Carter et al., 1980a). Although there are numerous means of divination employed in the Andes, divination with coca leaves is the most common and most respected (Carter et al., 1980a).

Diviners fall into many different categories according to their specialties and abilities and are known by an assortment of native names. "Yatiri," meaning "one who knows," is probably the most widespread term in both Quechua and Aymara. Many diviners have congenital deformities or have been (or claim to have been) struck by lightning (Carter et al., 1980a).

The act of divining or "reading" coca leaves takes many forms. It may be a formal ceremony performed by a specialist or an informal or personal act performed by an individual coquero. Indians who chew coca are intensely aware of the signs latent in the leaves they chew: in their form and color, in the taste and form of the chewed quid, or in the saliva which issues from it.

Formal divination involves the consultation of a knowledgeable yatiri at a specific time and place. A special woven cloth, the "cocatari" (Aymara) or "uncuña" (Quechua), is placed on the ground. A small handful of selected leaves is allowed to drop upon the cloth. The reading of the leaves depends upon many features of the leaves, including their color, shape, size, deformities, spots, holes, and creases as well as their spatial relationship to one another. Depending on all these factors, the leaves will symbolize death, bad or good luck, money, evil spells, a safe journey, or other things or will suggest the diagnosis or cure of an illness (Contreras Hernández, 1972; Frisancho Pineda, 1973; Carter et al., 1980a).

According to Martin (1970), diviners among the Incas would chew coca leaves and spit the juice into their palms with the two longest fingers extended. If the juice ran down both fingers equally, it was a good sign; if it ran down unequally, it was a bad one. Other diviners would burn coca leaves with llama fat and observe the way they burned.

Among the Campa of eastern Peru, coca is used by the shaman to determine the perpetrator of witchcraft. The shaman spits coca into his hand, shakes it, and ascertains the guilty party through its configuration (Ordinaire, 1892). The neighboring Machiguenga of the Peruvian montaña carve small idols out of coca wood. They believe that coca comes from benevolent spirits called "saanka'riite" and that it has the ability to reveal the future. For example, if a man chews coca and does not taste its sweetness, it is a sign of impending misfortune (Baer, 1970). Coca is also used in divination among tribes of the Northwest Amazon who use coca in powdered form. Future events may be foreseen by blowing a spoonful of coca powder into the air and observing the way it falls to the ground.

In Colombia, the Paez Indians of Cauca Department also employ coca in divination (Uscátegui, 1954), as did Chibcha priests in the central highlands at the time of the Spanish Conquest (Martin, 1970).

To summarize the importance of coca in Indian life, I would like to quote the eloquent remarks of Wagner (1978: 878): ". . . 'to chew coca' is part of the process through which the Quechuas absorb the depth of their culture and learn to understand what it means to be a *Runa,* a participant in traditional Quechua culture" (translation from Spanish mine) and of Martin (1970: 424): "Only appreciating the use of coca from the point of view of the Indian's cultural heritage, their beliefs and the necessities of their daily lives can give a proper perspective on the meaning of coca to these people."

Acknowledgments

Early phases of the research reported here were conducted at the Botanical Museum of Harvard University under a contract with the U.S. Department of Agriculture (no. 12-14-1001-230, R. E. Schultes, principal investigator). Financial support was also provided by a grant from the National Institute of Drug Abuse (no. 5 RO DA02110-02). I am grateful to the Field Museum Library staff for locating obscure references and to Ron Testa and Fleur Hales for preparing most of the photographic prints. I would like to acknowledge the Lowie Museum of Anthropology, University of California, Berkeley for supplying photographs of archeological coca from their collections and the Harvard Botanical Museum for allowing me to photograph coca-related artifacts from their exhibits. Geoffrey W. Conrad of the Peabody Museum of Archaeology and Ethnology supplied needed information about archeological materials at that institution. I am also grateful to the following persons who lent photographs: Robert Feldman, Laurent Rivier, Richard Evans Schultes and Oscar Tovar. Christine Niezgoda, Penny Matekaitis, and Roberta Becker helped immeasurably with the preparation of the manuscript. Phillip Rury and Laurent Rivier offered constructive criticism and suggestions for improving the text, for which I am most grateful.

Literature Cited

Albo, X. 1978. El mundo de la coca en Coripata, Bolivia. América Indígena 38: 938–969.

Antonil. 1978. Mama coca. Hassle Free Press, London.

Baer, G. 1970. Reise und Forschung in Ost-Peru. Verh. Naturf. Ges. Basel 80: 367–369.

Baker, P. T. & R. B. Mazess. 1963. Calcium: Unusual sources in the highland Peruvian diet. Science 142: 1466–1467.

Bejarano, J. 1945. El cocaísmo en Colombia. América Indígena 5: 11–20.

Billip, J. 1979. Coca chewers of Santa Marta. High Times, July: 69–73.

Bohm, B. A., F. R. Ganders & T. Plowman. 1982. Biosystematics and evolution of cultivated coca (Erythroxylaceae). Syst. Bot. 7: 121–133.

——, **D. W. Phillips & F. R. Ganders.** 1981. Flavonoids of *Erythroxylum rufum* and *Erythroxylum ulei*. J. Nat. Prod. 44: 676–679.

Bolton, R. 1976. Andean coca chewing: A metabolic perspective. American Anthropologist 78: 630–634.

Bray, W. 1978. The gold of El Dorado. Times Newspapers Ltd., London.

—— **& C. Dollery.** 1983. Coca chewing and high-altitude stress: A spurious correlation. Current Anthropology 24: 269–282.

Bües, C. 1911. La coca: Apuntes sobre la planta, su cultivo, beneficio, enfermedades y aplicación. Ministerio de Fomento. Imprenta Americana, Lima.

Burchard, R. E. 1974. Coca y trueque de alimentos. Pages 209–251. *In:* G. Alberti & E. Mayer, editors. Reciprocidad e intercambio en los andes peruanos. Instituto de Estudios Peruanos, Lima.

——. 1975. Coca chewing: A new perspective. Pages 463–484. *In:* V. Rubin, editor. Cannabis and culture. Mouton Publishers, The Hague.

——. 1976. Myths of the sacred leaf: ecological perspectives on coca and peasant biocultural adaptation in Peru. Doctoral dissertation, Indiana University, Bloomington.

Califano, M. & A. Fernández Distel. 1978. El empleo de la coca entre los Mashco de la Amazonia del Perú. Årstryck 1977. Göteborgs Ethnografiska Museum. pp. 16–22.

Carter, W. E., M. Mamani P. & J. V. Morales. 1981. Medicinal uses of coca in Bolivia. Pages 119–149. *In:* J. W. Bastien & J. M. Donahue, editors. Health in the Andes. American Anthropological Association, Washington, D.C.

——, ——, ——, **& P. Parkerson.** 1980a. Coca in Bolivia. Research Report, National Institute of Drug Abuse. La Paz, Bolivia.

——, **P. Parkerson & M. Mamani P.** 1980b. Traditional and changing patterns of coca use in Bolivia. Pages 159–164. *In:* F. R. Jerí, editor. Cocaine 1980. Pacific Press, Lima.

Castro de la Mata, R. 1981. La coca en la obra de Guaman Poma de Ayala. Bira (Lima) 11: 57–79.

Chávez Velásquez, N. A. 1977. La materia médica en el Incanato. Editorial J. Mejía Baca, Lima.

Ciuffardi T., E. N. 1948a. Dosis de alcaloides que ingieren los habituados a la coca. Revista de Farmacología y Medicina Experimental (Lima) 1: 81–99.

———. 1948b. Dosis de alcaloides que ingieren los habituados a la coca: nuevas observaciones. Revista de Farmacología y Medicina Experimental (Lima) 1: 216–231.

———. 1949. Contribución a la química del cocaísmo. Revista de Farmacología y Medicina Experimental (Lima) 2: 18–93.

Cohen, M. N. 1978. Archeological plant remains from the Central Coast of Peru. Ñawpa Pacha 16: 36–37.

Contreras Hernández, J. 1972. La adivinación por la coca en Chincheros (Perú). Proceedings of the 40th International Congress of Americanists (Rome) 2: 413–419.

Cruz Sánchez, G. & A. Guillén. 1948. Estudio químico de las substancias alcalinas auxiliares del cocaísmo. Revista de Farmacología y Medicina Experimental (Lima) 1: 209–215.

Daneri Pérez, M. R. 1974. El cultivo de la coca en el Perú. Tésis, Universidad Nacional Agraria La Molina, Lima.

Davis, E. W. 1983. The ethnobotany of chamairo: *Mussatia hyacinthina* (Bignoniaceae). J. Ethnopharmacol. 9: 225–236.

Dillehay, T. D. 1979. Pre-hispanic resource sharing in the central Andes. Science 204(6): 24–31.

Donnan, C. B. 1978. Moche art of Peru. Museum of Cultural History, University of California, Los Angeles.

Drolet, R. 1974. Coqueros and shamanism: An analysis of the Capulí Phase ceramic modeled figurines from the Ecuadorian Northern Highlands, South America. Journal of the Steward Anthropological Society 5(2): 99–121.

Duke, J. A., D. Aulik & T. Plowman. 1975. Nutritional value of coca. Bot. Mus. Leafl. 24: 113–119.

Engel, F. 1957. Early sites on the Peruvian coast. Southwestern Journal of Anthropology 13: 54–68.

———. 1963. A preceramic settlement on the central coast of Peru: Asia, Unit 1. Trans. Amer. Philos. Soc. 53(3): 77.

Fabrega, H. & P. K. Manning. 1972. Health maintenance among Peruvian peasants. Human Organization 31(3): 243–255.

Fine, N. L. 1960. Coca chewing: A social versus a nutritional interpretation. Unpublished manuscript. Columbia University, New York.

Frisancho Pineda, D. 1973. Medicina indígena y popular. Editorial J. Mejía Baca, Lima.

Gade, D. W. 1975. Plants, man and the land in the Vilcanota valley of Peru. W. Junk, The Hague.

Gagliano, J. A. 1960. A social history of coca in Peru. Doctoral dissertation, Georgetown University, Washington, D.C.

———. 1963. The coca debate in colonial Peru. The Americas 20(1): 43–63.

———. 1965. The popularization of Peruvian coca. Revista de Historia de América 59: 164–179.

———. 1968. Coca and environmental adaptation in the high Andes: An historical analysis of attitudes. Actas y Memorias, XXXVII Congreso Internacional de Americanistas, Buenos Aires 4: 227–236.

———. 1979. Coca and popular medicine in Peru: An historical analysis of attitudes. Pages 39–54. *In:* D. L. Browman & R. A. Schwarz, editors. Spirits, shamans, and stars: Perspectives from South America. Mouton Publishers, The Hague.

Ganders, F. R. 1979. Heterostyly in *Erythroxylum coca* (Erythroxylaceae). J. Linn. Soc., Bot. 78: 11–20.

Gentner, W. A. 1972. The genus *Erythroxylum* in Colombia. Cespedesia 1: 481–554.

Gifford, D. & P. Hoggarth. 1976. Carnival and coca leaf: Some traditions of the Peruvian Quechua Ayllu. Scottish Academic Press, Edinburgh.

Gosse, L.-A. 1861. Monographie sur l'*Erythroxylum coca*. Mem. Acad. Roy. Sci. Belgique 12(3): 1–145.

Grinspoon, L. & J. B. Bakalar. 1976. Cocaine: A drug and its social evolution. Basic Books, New York.

—— & ——. 1981. Coca and cocaine as medicines: An historical review. J. Ethnopharmacol. 3: 149–159.

Gutiérrez Noriega, C. & V. Zapata-Ortíz. 1948. Observaciones fisiológicas y patológicas en sujetos habituados a la coca. Revista de Farmacología e Medicina Experimental (Lima) 1: 1–31.

Hanna, J. M. 1976. Drug use. Pages 363–378. *In:* P. T. Baker & M. A. Little, editors. Man in the Andes. Dowden, Hutchinson & Ross, Stroudsburg, Pennsylvania.

Hegnauer, R. 1981. Chemotaxonomy of Erythroxylaceae (including some ethnobotanical notes on Old World species). J. Ethnopharmacol. 3: 279–292.

—— & L. K. Fikenscher. 1960. Untersuchungen mit *Erythroxylum coca* Lam. Pharm. Acta Helv. 35: 43–64.

Hemming, J. 1978. The search for El Dorado. Michael Joseph, London.

Hesse, O. 1891. A study of coca leaves and their alkaloids. Pharm. J. Trans. 21: 1109–1117, 1129–1135.

Holmstedt, B., E. Jäätmaa, K. Leander & T. Plowman. 1977. Determination of cocaine in some South American species of *Erythroxylum* using mass fragmentography. Phytochemistry 16: 1753–1755.

——, J.-E. Lindgren, L. Rivier & T. Plowman. 1979. Cocaine in blood of coca chewers. J. Ethnopharmacol. 1: 69–78.

Hulshof, J. 1978. La coca en la medicina tradicional andina. América Indígena 38: 837–846.

Javaid, J. I., M. W. Fischman, C. R. Schuster, H. Dekirmenjian & J. M. Davis. 1978. Cocaine plasma concentration: Relation to physiological and subjective effects in humans. Science 202: 227–228.

Jerí, F. R., editor. 1980. Cocaine 1980. Pacific Press, Lima.

Jones, J. 1974. Rituals of euphoria: Coca in South America. Museum of Primitive Art, New York.

Klepinger, L. L., J. K. Kuhn & J. Thomas, Jr. 1977. Prehistoric dental calculus gives evidence for coca in early coastal Ecuador. Nature 269: 506–507.

Kuczynski-Godard, M. H. & C. E. Paz Soldán. 1948. Disección de indigenismo peruano. Un examen sociológico e médico-social. Publicaciones del Instituto de Medicina Social, Lima.

Kutscher, G. 1955. Arte antiguo de la costa norte del Perú. Gebrüder Mann, Berlin.

Lathrap, D. W., D. Collier & H. Chandra. 1976. Ancient Ecuador: Culture, clay and creativity, 3000–300 B.C. Field Museum of Natural History, Chicago, Illinois.

Le Cointe, P. 1934. A Amazônia Brasileira III: Árvores e plantas utéis. Livraria Clássica, Belém, Brazil.

Lothrop, S. K. 1937. Coclé: An archaeological study of central Panama. Memoirs of the Peabody Museum of Archaeology and Ethnology, Harvard University 7: 1–206.

Lyons, A. B. 1885. Notes on the alkaloids of coca leaves. Amer. J. Pharm. 57: 465–477.

Machado C., E. 1972. El género *Erythroxylon* en el Perú. Raymondiana 5: 5–101.

——. 1980. Determination of varieties and cultivars in Peruvian coca. Pages 239–245. *In:* F. R. Jerí, editor. Cocaine 1980. Pacific Press, Lima.

Mariani, A. 1890. Coca and its therapeutic application. J. N. Jaros, New York.

Martin, R. T. 1970. The role of coca in the history, religion and medicine of South American Indians. Econ. Bot. 24: 422–438.

Mayer, E. 1978. El uso social de la coca en el mundo andino: Contribución a un debate y toma de posición. América Indígena 38: 849–865.

Meggers, B. J. 1966. Ecuador. Praeger, New York.

Miner, H. 1939. Parallelism in alkaloid-alkali quids. Amer. Anthropol. 41: 617–619.

Montesinos, F. A. 1965. Metabolism of cocaine. Bulletin on Narcotics 17: 11–19.

Morris, D. 1889. Coca. Bull. Misc. Inform. 25: 1–13.

Mortimer, W. G. 1901. History of coca. J. H. Vail, New York.

Moser, B. & D. Taylor. 1965. The cocaine eaters. Taplinger Publishing Co., New York.

Naranjo, P. 1974. El cocaísmo entre los aborígenes de Sud América: Su difusión y extinción en el Ecuador. América Indígena 34: 605–628.

Ochiai, I. 1978. El contexto cultural de la coca entre los indios Kogi. América Indígena 38: 43–49.

Ordinaire, O. 1892. Du Pacifique a l'Atlantique par les Andes Péruviennes et l'Amazone. Plon, Nourrit & Co., Paris.

Paly, D., P. Jatlow, C. Van Dyke, F. Cabieses & R. Byck. 1980. Plasma levels of cocaine in native Peruvian coca chewers. Pages 86–89. *In:* F. R. Jerí, editor. Cocaine 1980. Pacific Press, Lima.

Patiño, V. M. 1967. Plantas cultivadas y animales domésticos en América equinoctial. Vol. 3: Fibras, medicinas, misceláneas. Cali, Colombia.

Patterson, T. C. 1971. Central Peru: Its population and economy. Archaeology 24: 316–321.

Peña Begué, R. de la. 1972. El uso de la coca entre los Incas. Revista Española de Antropología Americana 7: 277–305.

Pérez de Barradas, J. 1940. Antigüedad del uso de la coca en Colombia. Revista Acad. Colomb. Ci. Exact. 3: 323–326.

Pickersgill, B. & C. B. Heiser, Jr. 1976. Cytogenetics and evolutionary change under domestication. Philos. Trans., Ser. B 275: 55–69.

Picón-Reátegui, E. 1976. Nutrition. Pages 233–234. *In:* P. T. Baker & M. A. Little, editors. Man in the Andes. Dowden, Hutchinson & Ross, Stroudsburg, Pennsylvania.

Plowman, T. 1979a. Botanical perspectives on coca. J. Psychedelic Drugs 11: 103–117.

———. 1979b. The identity of Amazonian and Trujillo coca. Bot. Mus. Leafl. 27: 45–68.

———. 1980. Chamairo: *Mussatia hyacinthina*—An admixture to coca from Amazonian Peru and Bolivia. Bot. Mus. Leafl. 28: 253–261.

———. 1981. Amazonian coca. J. Ethnopharmacol. 3: 195–225.

———. 1982. The identification of coca (*Erythroxylum* species): 1860–1910. J. Linn. Soc., Bot. 84: 329–353.

———. 1984. The origin, evolution and diffusion of coca (*Erythroxylum* spp.) in South and Central America. *In:* D. Stone, editor. Pre-Columbian plant migration. Papers of the Peabody Museum of Archaeology and Ethnology, Vol. 76. Harvard University, Cambridge, Massachusetts.

——— & L. Rivier. 1983. Cocaine and cinnamoylcocaine content of thirty-one species of *Erythroxylum* (Erythroxylaceae). Ann. Bot. (London) 51: 641–659.

——— & A. T. Weil. 1979. Coca pests and pesticides. J. Ethnopharmacol. 1: 263–278.

Prance, G. T. 1972. Ethnobotanical notes from Amazonian Brazil. Econ. Bot. 26: 228–232.

Quijada Jara, S. 1950. La coca en las costumbres indígenas. Published by the author, Huancayo, Peru.

Reens, E. 1919a. La coca de Java: Monographie historique, botanique, chimique et pharmacologique. Lucien Declume, Lons-le-Saunier, France.

———. 1919b. La coca de Java. Bull. Sci. Pharmacol. 21: 497–505.

Reichel-Dolmatoff, G. 1950. Los Kogi: Una tribu de la Sierra Nevada de Santa Marta, Colombia. Revista del Instituto Etnológico Nacional 4: 75–79.

———. 1953. Contactos y cambios culturales en la Sierra Nevada de Santa Marta. Revista Colombiana de Antropología 1: 17–122.

———. 1972. San Agustín. Praeger Publishers, New York.

Rivier, L. 1981. Analysis of alkaloids in leaves of cultivated *Erythroxylum* and characterization of alkaline substances used during coca chewing. J. Ethnopharmacol. 3: 313–335.

Romburgh, P. van. 1894. Sur quelques principes volatils des feuilles de coca cultivées a Java. Recueil Trav. Chim. Pays-Bas Belg. 13: 425–428.

———. 1895. Over eenige vluchtige bestanddeelen van de op Java gekweekte Cocabladen. Verslagen Zittingen Wis-Natuurk. Afd. 3: 181–183.

Rostworowski, M. de Diez Canseco. 1973. Plantaciones prehispánicas de coca en el vertiente del Pacífico. Revista del Museo Nacional (Lima) 39: 193–224.

Rury, R. P. 1981. Systematic anatomy of *Erythroxylum* P. Browne: Practical and evolutionary implications for the cultivated cocas. J. Ethnopharmacol. 3: 229–263.

———. 1982. Systematic anatomy of the Erythroxylaceae. Ph.D. dissertation, University of North Carolina, Chapel Hill.

——— & **T. Plowman.** 1984. Morphological studies of archeological and recent coca leaves (*Erythroxylum* spp., Erythroxylaceae). Bot. Mus. Leafl. 29: 297–341.

Rusby, H. H. 1888. Coca at home and abroad. Therap. Gaz. 12: 158–165, 303–307.

Saenz, L. N. 1941. El coqueo factor de hiponutrición. Revista de la Sanidad de Policía (Lima) 1: 129–147.

Schultes, R. E. 1957. A new method of coca preparation in the Colombian Amazon. Bot. Mus. Leafl. 17: 241–246.

———. 1981. Coca in the northwest Amazon. J. Ethnopharmacol. 3: 173–194.

——— & **B. Holmstedt.** 1968. The vegetal ingredients of the myristicaceous snuffs of the northwest Amazon. Rhodora 70: 119.

Schulz, O. E. 1907. Erythroxylaceae. Das Pflanzenreich 4(134): 1–164.

South, R. B. 1977. Coca in Bolivia. Geogr. Rev. (New York) 67: 22–33.

Squibb, E. R. 1885. Coca at the source of supply. Pharm. J. Trans. 16: 46–49.

Stone, D. 1977. Pre-Columbian man in Costa Rica. Peabody Museum Press, Cambridge, Massachusetts.

Towle, M. A. 1961. The ethnobotany of pre-Columbian Peru. Wenner-Gren Foundation, New York.

Turner, C. E., M. A. Elsohly, L. Hanus & H. N. Elsohly. 1981a. Isolation of dihydrocuscohygrine from Peruvian coca leaves. Phytochemistry 20: 1403–1405.

———, **C. Y. Ma & M. A. Elsohly.** 1981b. Constituents of *Erythroxylon coca* II: Gas-chromatographic analysis of cocaine and other alkaloids in coca leaves. J. Ethnopharmacol. 3: 293–298.

United Nations. 1980. Statistics on narcotic drugs for 1978. International Narcotics Control Board, Vienna & New York.

Uscátegui, N. 1954. Contribución al estudio de la masticación de las hojas de coca. Revista Colombiana de Antropología 3: 207–289.

———. 1961. Distribución actual de las plantas narcóticas y estimulantes usadas por las tribus indígenas de Colombia. Revista Acad. Colomb. Ci. Exact. 11: 215–228.

Wagner, C. A. 1978. Coca y estructura cultural en los Andes peruanos. América Indígena 38: 877–902.

Wavrin, M. de. 1937. Moeurs et coutumes des indiens sauvages de l'Amerique du Sud. Payot, Paris.

Weil, A. T. 1975. The green and the white. Pages 318–336. *In:* G. Andrews & D. Solomon, editors. The coca leaf and cocaine papers. Harcourt Brace Jovanovich, New York.

———. 1976. A gourmet coca taster's tour of Peru. High Times, May: 44–47, 76–81, 88–89.

————. 1981. The therapeutic value of coca in contemporary medicine. J. Ethnopharmacol. 3: 367–376.

Willaman, J. J. & B. G. Schubert. 1961. Alkaloid-bearing plants and their contained alkaloids. Technical Bulletin 1234. Agricultural Research Service, U.S. Department of Agriculture, Washington, D.C.

Yacovleff, E. & F. L. Herrera. 1934–1935. El mundo vegetal de los antiguos peruanos. Revista del Museo Nacional (Lima) 3: 243–326; 4: 31–102.

Zapata-Ortíz, V. 1970. The chewing of coca leaves in Peru. International Journal of Addiction 5: 287–294.

A Preliminary Report on Diversified Management of Tropical Forest by the Kayapó Indians of the Brazilian Amazon

Darrell A. Posey

Passionate journalistic descriptions and detailed scientific studies of the disasters wrought by "development" on the Amazon Basin have served to focus world concern not only on the loss of Amazonia but also on the destruction of tropical forests throughout the planet. The so-called "dilemma" of modern development in frontier areas is, of course, how to provide a framework for exploitation of natural resources without the usual ecological destruction and social inequities that characterize current practices (Barbira-Scazzocchio, 1980; Moran, 1983).

"New" ideas for development seem to come from the same old sources: plopped-down technology and snatched-up varieties of "improved" cultivars that result in large monocultural fields that are designed to accommodate the latest machines from distant factories and that survive on life-support systems of fertilizers and pesticides. In prominent journals scientists tell us that Amazonian soils can be continuously cropped just like those at their experimental plots—with only the aid of these chemical "uppers!" They fail to point out, analyze, or otherwise consider the social and ecological effects of such chemicals in a region as vast as Amazonia. Such limited thinking has already led to disruptive shifts in land tenure (Barbira-Scazzocchio, 1980: iii–xiv); for example, Hecht (1983: 177) reports that 55% of the private lands of Amazonia is now controlled by 3.3% of the landowning population. This shift is the epitomy of the Amazonian problem: peasants, small farmers, and Indians simply cannot afford the expenses of agricultural chemicals any more than the ecological system can accommodate large-scale deforestation and monocultures. The world seems doomed to a long series of "green revolutions" turned brown by wholesale transplanting of technology without concern for social consequences. It is unfortunate, but tragically true, that blindly specialized scientists are one of the greatest threats to developing regions.

In recent years, a small cadre of ethnobiologists has set out to compensate for the ethnocentrism of Western science by investigating the sophistication of native concepts of ecology and knowledge of natural resources. Ethnobiological data offer an untapped wealth of information about the biological diversity of regions like Amazonia. Alternative models of development based upon this indigenous and folk knowledge have been proposed as ecologically sound and socially progressive paths out of the current developmental dead ends (Posey, 1983a; Posey et al., in press). Several projects such as those being conducted by the New York Botanical Garden, the Instituto Nacional de Pesquisas da Amazônia-Manaus, and the Laboratório de Etnobiologia (São Luís, Maranhão) are systematically collecting ethnobiological information from indigenous peoples. This data base can be used to formulate new strategies of resource exploitation which are consistent with indigenous ecological models and based upon long-range management of native plants and animals.

Progress, however, is slow; research funds have been difficult to obtain because ethnobiology (and its various subfields) fall outside the conservative paradigm of established scientific research. Another fundamental problem is finding scientists who are of a personal ilk that allows them to shed the mantle of Western scientific superiority to learn from "primitives." Scientists accustomed to working with plants and animals often view work with "cultural consultants" as complicating and "style-cramping." People cannot be clipped, pressed, and put away depending upon weather and mood. Of equal difficulty, such consultants often do not know how to respond to questions framed in Western terminology and logic, while scientists rarely take the time to learn enough of the language and culture under study to ask "proper questions" framed in non-Western logic. Nonetheless, great strides have been made to put the ethnobiological sciences on firm footing.

One of the areas in which I feel ethnobiology in general, and ethnobotany in particular, can make a significant contribution to research in tropical regions is through the study of indigenous use and management of tropical forests. Increasing evidence of large, dense aboriginal populations in Amazonia (Denevan, 1976; Dobyns, 1966; Smith, 1980) substantiates that reliance on forest resources was much more important than previously assumed; thus natural resources in the forest are of greater potential and variety than realized. Rapid deforestation of primary tropical forests around the world also makes knowledge of secondary forests of greater importance for the future because secondary forests now are the dominant forest type. This paper deals with the importance and use of Amazonian forest by the Kayapó Indians of the Brazilian Amazon. When botanical identification are completed by appropriate specialists, more precise details and examples will be available.

The Kayapó

The Kayapó, one of the most populous (ca. 2500 people) of the Jê tribes, live on a two-million-hectare reserve in the Brazilian state of Pará (Fig. 1). Although the reserve is supposedly to be protected by official government demarcation, the Kayapó are effectively under siege by gold miners ("garimpeiros"), large plantation owners ("fazendeiros"), and settlers ("pouseiros"). Economic and social changes are rapidly destroying traditional Kayapó culture; it is tragic to see millenia of accumulated experience and knowledge disappear in a few generations with hardly a lamentation by the engulfing society.

Our current interdisciplinary ethnobiological project is attempting to record the wealth of folk ecological knowledge still possessed by the Kayapó. Some of the most important data relevant to alternative models of Amazonian development come from the following categories of secondary forest use.

Management and use of secondary forest

"ANYTHING-BUT-ABANDONED FIELDS"

One of the most persistent scientific myths about indigenous slash/burn agriculture is that fields are abandoned within a few years after initial clearing and planting. Studies of Indian agricultural systems have recently detailed the long-range use of old fields by Amazonian Indians, showing that fields are anything but abandoned (William Denevan, pers. comm.). In fact, old fields offer an important concentration of highly diverse natural resources, including medicinal and food plants as well as game animals (Alcorn, 1981).

Kayapó "new fields" ("puru nu") peak in production of principal cultivars in two or three years but continue to bear domesticated plant products for many years (sweet potatoes for four to five years, yams and taro for five to six years, manioc for four to six years, and papaya for five or more years). Some banana varieties continue to bear fruit for 15 to 20 years, "urucú" (*Bixa orellana*) for 25 years, and "cupá" (*Cissus gongylodes*) for 40 years. Cupá is an important Kayapó domesticate, little known outside the Indian area (Kerr et al., 1978). The Kayapó consistently revisit old fields ("puru tum") throughout the area, especially those near abandoned village sites, seeking these lingering riches.

Old fields take on new life, however, as the myriad of plants in the natural reforestation sequence begin to appear, creating a type of forest ("bà-ràràrà") in which light penetrates to the forest floor. These plants (of which a representative list is found in Table I) provide a wide range of useful products, including: food and medicine, fish and bird baits, thatch, packaging, paints, oils, insect repellents, firewood, construction materials, fibers for ropes and cords, body cleansers, and products for craft production—to name but a few.

Old fields are perhaps most important for their concentrations of medicinal plants. In our recent survey of plants found in "puru tum," 94 percent of the 368 plants collected were of medicinal significance!

Another function of old fields is the attraction of wildlife to the food provided by the abundant, low, leafy plants (Table I). High forests, in contrast, provide little such food and consequently are sparse in game. Intentional dispersal throughout the area of old fields and managment of these by systematic hunting extends, therefore, the human influence over the forest by providing, in effect, large "game farms" near human population concentrations (Posey, 1982). A delicate balance is necessary to manage these old fields. Small plants are intended to attract game animals, but the game population cannot get too dense or severe damage to crops will result. In the Kayapó division of labor, the women work in the fields while the men hunt nearby in the forest zone that surrounds their wives' fields. This hunting not only provides meat for food but also protects new fields from excessive animal destruction.

Game animals are particularly attracted to fruit trees. The Kayapó plant fruit trees (Table II) in new and old fields, as well as along trails, with an eye to a future that will provide products ripening throughout the year. Tree plantings illustrate long-term planning and forest management since many of the trees require decades to bear fruit; "castanha do Pará" (Brazil nut), for example, does not produce its

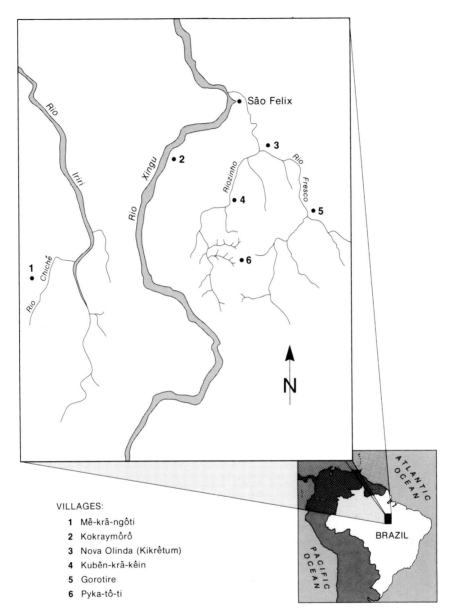

VILLAGES:
1 Mẽ-krã-ngôti
2 Kokraymôrô
3 Nova Olinda (Kikrêtum)
4 Kubẽn-krã-kêin
5 Gorotire
6 Pyka-tô-ti

FIG. 1. Geographic position of the Kayapó Indian villages of central Brazil.

first nuts for 25 years. Most of these trees also produce food favored by the Indians and are an essential part of Kayapó subsistence. These old fields should perhaps be called "game-farm orchards" to emphasize their diverse resources (Smith, 1977; Posey et al., in press).

SEMI-DOMESTICATES IN OLD FIELDS

Old fields also serve as an important repository of "semi-domesticated" plants. The exact terminology of this category of plant is yet to be worked out. I use the term "semi-domesticate" to indicate plants that are intentionally manipulated by

Table I

Some plant-soil-animal associations in the selected ecozone (bà-ràràrà)[a]

| Plant species | Use of plant | | Kayapó name | Soil type[b] | Associated animals[c] |
	By man	By animal			
Humiria balsamifera (Aubl.) St. Hil.	eat fruit	eat fruit	bà-rerek	1, 2	A, B, C, D, E
Psidium guyanense Pers.	eat fruit	eat fruit & leaves	kamokàtytx	1, 2	F
Zingiberaceae	use root for tea; smoke leaves	eat leaves	màdn-tu	3	F
Peschiera sp.	use for paint		pita-teka	3	
Catasetum sp.	medicinal		pitu	2, 3	
Bignoniaceae	medicinal	eat leaves	ngra-kanê	2, 3	C, F
Cissampelos sp.	fish bait	eat fruit	tep-kanê	1, 2, 3	C, D
Piperaceae	fish bait	eat fruit	makrê-kanê	1	A, B, C, D
Amasonia sp.	prophylaxis		pidjó-rã	3	
Oenocarpus distichus Mart.	eat fruit	eat fruit	kamêrê (bacaba)	1	A, B, C, D
Monotagma sp.	grind leaves, eat roots	eat leaves, eat roots	kùryre	1, 3	F
Myrcia sp.	eat fruit	eat fruit & leaves	kônhôkô	1, 2	A, C, D, F
Cecropia leucocoma Miq.		eat fruit	atwỳrà'ô'	1, 3	H, F
Polypodiaceae	medicinal		tôn-kanê	2	
Clarisia ilicifolia (Spreng.) Lang. & Rossb.	medicinal	eat leaves	pidjô-nire	2	F
Centrosema carajasense Cavalc.	fish poison		akrô	2, 3	
Cassia hoffmanseggii Mart. ex Benth.	medicinal	eat fruit & leaves	pidjô-kakrit	2, 3	C, D, F

[a] Identifications of plants made by Dr. Susanna Hecht, Department of Geography, University of California, Los Angeles.

[b] 1 = black ("pyka-tyk"), 2 = red ("pyka-kamrek"), 3 = yellow ("pyka-ti").

[c] A = white-lipped peccary ("porção"), B = white paca ("paca branca"), C = agouti ("cutia branca"), D = tortoise ("jaboti"), E = red paca ("paca vermelha"), F = red agouti ("cutia vermelha"), G = deer ("veado"), H = tapir ("anta").

the Indians, who modify the plant's habitat to stimulate growth. The genetic consequences of this process are still unknown, but are under study as part of our Kayapó project. There are two categories of these semi-domesticates: primary forest transplants and secondary forest transplants.

Primary forest transplants.—Ethnoecological classification of forest types by the Kayapó is complicated and too detailed for discussion here (but see Posey, 1983a). Relatively open forests are given special names ("bà-ràràrà" and "bà-epti") and are known refuges for light-loving plants that also grow well in old fields ("puru tum"). Gathering trips to the primary forest are frequently combined with hunting expeditions, although specific trips may be made solely to gather certain plants from the forest for transplanting into old fields. Concentrations of semi-domesticates in old fields minimize both the indigenous effort necessary to locate needed plants and the lack of natural forest resources.

Secondary forest transplants.—In areas where the forest is disturbed by either natural or man-made events, habitats are created that in essence replicate field clearings. For example, large forest openings caused by trees that have fallen through natural processes (old age and storms) or that have been felled by Indians to raid bee hives, create microenvironmental conditions similar to those of field clearings (Posey, 1982, 1983b). Likewise, openings due to abandoned camps and village sites or wide swaths left by trails are also reserves for plants that thrive in old fields. These areas are visited on gathering trips with the goal of transplanting certain secondary forest plants into old fields closer to villages, thereby making secondary forest products more readily available.

"FOREST-FIELDS"

The Kayapó custom of transplanting is only part of a much broader system that I have described (Posey, 1983a) as "nomadic agriculture" and that was undoubtedly widespread in other Amazonian tribes. Until recently, Kayapó groups travelled extensively in the vast areas between the east-west boundaries of the Tocantins and Araguaya Rivers and the north-south limits of the Planalto and the Amazon River (Turner, 1966). Remnants of this once extensive system give hints of the importance of "forest-fields" ("pry-jakrê") to aboriginal subsistence. Today the Kayapó still carry out several month-long treks per year, although much of the old network of trails and campsites is now abandoned.

Food and utensils are not carried by the Indians on treks because of their bulk and weight. Food gathering for 150 to 200 people cannot, however, be left solely to chance. Thus gathered plants are transplanted to concentrated spots near trails and campsites. The "forest fields" make readily available to future passersby the necessities of life, including food, medicinals, cleansing agents, hair and body oils, insect repellents, leaves for cooking, house construction materials, etc.

Forest-fields intentionally replicate naturally occurring "resource islands" (e.g., "pidjô kô"), which are areas in the primary forest where specific concentrations of useful plants or animals are found. Figure 2 shows naturally occurring resource islands between the modern Kayapó village of Kubenkrãkêin and the abandoned village of Pykatôti. These resource islands include: hunting zones, fish concentrations, palmito and palm nut sources, cane for arrows, etc.

Dependency on naturally occurring "resource islands" and their man-made "forest-field" counterparts allows the Kayapó groups to travel months at a time without need of domesticated garden produce.

TRAILSIDE PLANTINGS

In addition to the "forest-fields" near campsites and alongside Kayapó trails, the sides of trails ("pry kôt") themselves are planting zones (Fig. 3). It is not uncommon to find Kayapó trails composed of four-meter-wide cleared strips of forest. It is difficult to estimate the extensiveness of the aboriginal trails that interconnected distant Kayapó villages (Fig. 1). A conservative estimate of existing

Table II
Tree species planted by the Kayapó Indians[a]

Scientific name	Portuguese name	Kayapó name	Planted for		Attracts	
			Food	Other use	Game	Fish
Alibertia edulis A. Rich.	marmelada (lisa)	motu	X		X	
Alibertia sp.	marmelada do campo	roi-krāti	X		X	
Anonna crassiflora Mart.	araticum	ongrê	X			
Artocarpus integrifolia L. f.	jacá	jacá	X			
Astrocaryum tucuma Mart.	tucum (2 varieties)	roi-ti (mrà)	X	salt		
Astrocaryum vulgare Mart.	tucumã	woti	X	oil		
Bertholletia excelsa Humb. & Bonpl.	castanha do Pará	pi'ỳ	X			
Bixa orellana L.	urucú (4 varieties)	pỳ kumrenx, pỳ poi ti, pỳ krã re, pỳ ja-biê		body paint		
Byrsonima crassifolia H.B.K.	muruci	kutenk	X			
Caryocar villosum (Aubl.) Pers.	piqui (3 varieties)	prĩ kā ti	X		X	
		prĩ krã ti	X			
		prĩ kumrenx	X			
Citrus aurantifolia (Christm.) Swingle)	lima	pidjô ngrã ngrã	X			
Citrus aurantium L.	laranja	pidjô ti	X			
Citrus limonia Osbeck	limão	pidjô poi re	X			
Coffea arabica L.	café	kapê	X			
Cordia sp.	cereja Kayapó	kudjã redjô	X		X	
Endopleura uchii (Huber) Cua-trec.	uxi	kremp	X			

Table II (continued)

Scientific name	Portuguese name	Kayapó name	Planted for		Attracts	
			Food	Other use	Game	Fish
Eugenia jambos L.	jambos	pidjô nore	X		X	
Euterpe oleracea Mart.	açai (2 varieties)	kamere kàk, kamere kàk ti	X		X	
Genipa americana L.	genipapo (2 varieties)	mroti, mrotire	X	body paint		
Hancornia speciosa Gomez	mangaba	pi-ô-tire	X			
Hymenaea courbaril L.	jatobà	moi (motx)	X		X	
Inga sp.	inga (6 varieties)	kohnjô-kô (jaka, kryre, poire, tire, ngrãngrã, tyk)	X		X	
Lecythis usitata Miers	sapucaia	kromu	X			
Lecythis usitata var. paraensis (Ducke) Knuth	sapucaia	pi'y tê krê ti	X			
Mangifera indica L.	manga	kuben poi re	X			
Manilkara huberi (Ducke) Stand.	massaranduba	krwyà no kamrek			X	X
Mauritia martiana Spruce	buritirana	ngrwa râre	X			
Mauritia vinifera Mart.	buruti	ngrwa	X			
Maximiliana regia Mart.	inajá	rikre	X	salt		
Oenocarpus bacaba Mart.	bacaba	kamere	X		X	
Orbignya eichleri Drude	piaçaba	ngra djàre	X			
Orbignya martiana Barb. Rodr.	babassu	rõ	X	oil, salt		
Parinari montana Aubl.	pariri	kamô	X		X	
Persea americana Mill.	abacate	kaprã	X			
Platonia insignis Mart.	bacuri	pî panhê ka tire	X		X	

Table II (continued)

Scientific name	Portuguese name	Kayapó name	Planted for		Attracts	
			Food	Other use	Game	Fish
Pourouma cecropiaefolia Mart.	inbauba	atwỳrà krã krê	X		X	
Pouteria macrophylla (Lam.) Eyma	tuturubá	kamokô	X		X	
Psidium guajava L.	goiaba	pidjô kamrek	X			
Ravenala guyanensis Steud.	banana brava	tytyti djô	X			
Rollinia mucosa Baill.	biribá	biri	X			
Solanum paniculatum L.	jurubeba	miêchet ti	X		X	X
Spondias lutea L.	cajá		X			
Spondias lutea L. (S. mombim L.)	taperaba	bàrere-krã-kryre	X			
Theobroma cacao L.	cacau	kuben krã ti	X		X	
Theobroma grandiflorum K. Schum.	cupuaçu	bàri-djô	X			

ᵃ Identifications based upon Cavalcante (1972, 1974, 1979) from comparisons with common names of the region; systematic specimen collection is now underway.

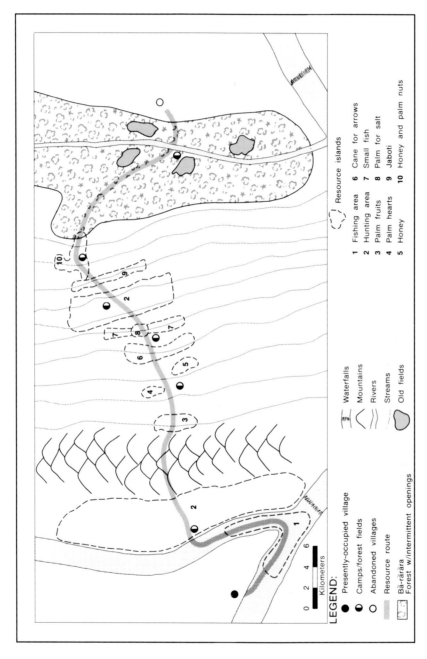

Fig. 2.   Route of trek from Kubẽn-krã-kéin to abandoned village site (Pyka-tó-ti) showing resource islands and campsites associated with forest fields.

trails associated with Gorotire (one of 11 modern Kayapó villages) yields 500 km of trails that average 2.5 meters wide. Thus, the total area under management is hardly insignificant. Trailsides are planted with numerous varieties of yams, sweet potatoes, Marantaceae, *Cissus,* Zingiberaceae, Araceae, Cannaceae, and other un-identified, edible, tuberous plants. Hundreds of medicinal plants and fruit trees also increase the diversity of the planted flora.

In a survey of a three-kilometer trail leading from Gorotire to a nearby garden, the following were observed: 1) 185 planted trees representing at least 15 different species, 2) approximately 1500 medicinal plants of an undetermined number of species, and 3) approximately 5500 food-producing plants of an undetermined number of species.

Not all trails are as heavily managed as this sample, but the importance of trailside planting and the extensiveness of this little-known ecological management system to aboriginal subsistence strategy merits intensive investigation.

The immediate one- to four-meter wide swath provided by trail clearing is *not* the entire effective distance of human activity. An additional factor is the distance away from the trail that the Kayapó choose for defecation/urination. I have measured the average distance, which is a rather culturally fixed proxemic unit, at five meters (or ten meters in width, considering both sides of the trail).

While squatting to defecate, the Kayapó often plant tubers, seeds, or nuts they have collected during the day and stored in a fiber pouch or bag. This activity, combined with the natural process of seed transportation through fecal material, makes the overall distance near trails under human influence even more extensive and significant. The effect is further accentuated by the age of the trails—some date to the ancient days of the ancestors.

PLANTATIONS IN FOREST OPENINGS

For the Kayapó, openings in the primary forest are called "bà krê ti" and are seen as natural prototypes for gardens. There are two types of "bà krê ti": 1) openings caused by trees that fall due to old age or storms, and 2) openings that are man-made by felling large trees to take honey from bees. Both types of forest openings create new micro-habitats and planting zones because the light penetrates to forest floor and creates conditions similar to those of garden plots. Although the Kayapó have a special name for their gardens ("puru"), they also use "bà krê ti" to describe gardens as well as "forest-fields." The idea for planting gardens may have come from their study and use of "bà krê ti" or may be a logical extension of their management of such forest openings.

Into "bà krê ti" are transplanted domesticates and semi-domesticates like va-rieties of manioc, taro, cupá, yams, sweet potatoes, taioba, beans, and arrowroot and related plants. These thrive in such habitats, and, according to Kayapó ag-riculturalists, their productivity is significantly increased.

The intriguing characteristics of this continuum of man-nature interaction high-lights the difficulty our science has with analyzing and evaluating the nebulous categories of man-made vs. natural and domesticate vs. non-domesticate. The entire process of domestication in relation to indigenous perceptions and man-agement of ecologically similar habitats is an important topic and one of the foci of our current research.

HILL GARDENS

Another form of agriculture that is related to "bà krê ti" plantations is the "krãi kam puru" or "hill garden." Tuberous plants, like Zingiberaceae, Araceae, and

FIG. 3. Bep-to-poop, a well-known shaman from the village of Gorotire, inspects a plantation of
"kwyrà re," an important medicinal. This plantation is found along a four-meter-wide trail in the
planting zone called "pry-kôt" (margin of the trail).

Marantaceae varieties, are planted in these well-drained, hillside plots. These fields
are principally reserved for food sources in case of floods or crop disasters and
are considered to be very valuable plant "banks." Hill gardens are exclusively
kept by old women ("mẽ-benget") under the direction of the Kayapó female chief
("menire nhõ benadjwỳrà"), the highest ranking female authority. These gardens
are apparently formed in two ways. In the first, old fields of eight to ten years of
fallow are cleared of underbrush. Pieces of tuber stock are planted in shallow
holes in fertile pockets of soil. Planting occurs when the new rains have soaked
the soils in September. Little care is required to maintain these fields. Harvest
occurs at the onset of the dry season (June), although representative plant varieties
are always left behind to preserve the tuber "bank."

The second type of hill garden is based upon management of plant communities
associated with bananas. As banana trees grow in maturing new fields, they produce shade, modify soil conditions, and, thus, produce a specialized microenvironment. The Kayapó know approximately two dozen varieties of edible tubers
and numerous medicinal plants that thrive under these conditions and are planted
near the banana trees ("tytyti kô"). This microenvironment thus becomes a planting zone within maturing and old fields. These plants, called "companions of the
banana" ("tytyti kotam"), continue to grow with the banana until the height of
secondary forest growth no longer is conducive to this plant community. When
this occurs, shoots of old bananas are transferred to new fields, while the "companions" are transplanted to already established plantations of bananas in other
maturing fields.

This type of hill garden illustrates not only how Indians exploit the properties
of fields in transition between new and old (in Kayapó, between "puru nu" and
"ibê tum") but also how microenvironmental planting zones are created to modify
effects of secondary forest growth. Equally significant is the indigenous concep-

FIG. 4. Uté shows a young babassu palm that he has planted in the "quintal" (backyard) of his house. Palms are widely planted in old fields and near village sites.

tualization of plant communities ("kotam"), rather than individual species, as the basis for ecological management. Other plant companions are under investigation for papaya, genipapa, and urucú, all of which are viewed as foci of other managed plant communities.

QUINTAL MANAGEMENT

"Quintal" is a Portuguese word that describes areas adjacent to homes that are generally planted with useful or decorative plants (Fig. 4). The idea is more ancient than the European introduction, since the Kayapó too rely on areas near their homes ("ki krê bum") to grow useful plants. To date, a partial survey has produced 86 varieties (estimate based on tentative identification) of food plants and dozens of additional medicinal plants.

The practice of medicine by the Kayapó is highly elaborate. Almost every household has its complement of common medicinal plants, many of which are domesticates or semi-domesticates. Shamans ("wayanga") specialize in different disease treatments, each of which requires specific plants. Dozens of "medicine knowers" ("pidjà mari") also effect minor cures with their own array of medicinal plants. Medicinal plants are often kept in secret forest plantations since their use forms part of the private knowledge of the curer; others are overtly grown in the quintal and only their use is secret. Thus each quintal reflects the medicinal knowledge and specialization (or lack thereof) of its occupants.

A major, long-term result of quintal management is the formation of topsoil. Some of the richest and most productive soils in Amazonia are those called "terra preta dos índios," believed to have been produced by Indian manipulation of generally poor Amazonian soils (Smith, 1980). This process, as yet undescribed, will be investigated in an upcoming phase of the Kayapó project.

Conclusion

Indigenous use and management of tropical forests are best viewed as continua between plants that are domesticated and those that are semi-domesticated, manipulated, or wild. Likewise, there is no clearcut demarcation between natural and managed forest: much of what has been considered "natural" forest in Amazonia is probably the result of millenia of human management and co-evolution. The Kayapó system of resource management utilizing forest-fields, hill gardens, trail-side plantings, and old fields to concentrate useful plants and animals underlines the inadequacies of the narrow paradigms that restrict Western science.

Preliminary analyses of data collected in the Projeto Kayapó offer a wealth of information about useful plants in the tropical forest, particularly plants that occur in, readily adapt to, or are stimulated by transplanting into secondary forest planting zones. New insights into the processes of plant domestication have also been found as well as new ideas for long-range ecological management of tropical forests that include human exploitation of integrated plant and animal communities. In brief, indigenous management strategies dwarf the current developmental practices that involve short-term gain without consideration of long-term social and environmental costs.

Ethnobiological research can provide new avenues for research in the Amazon Basin—or wherever indigenous and folk societies survive. Money, scientists willing to learn from other cultures, and, of course, time are lacking. Indigenous cultures are becoming extinct by the day! We must not only work furiously to record valuable data, but we must fight to preserve the lands, freedom, and cultures of native peoples. They are a living, human wealth whose loss our planet can ill afford.

Acknowledgments

This is a preliminary report based upon research conducted in 1977–79, with plant collections made in January 1983 with the aid of Dr. Gerhard Gottsberger, now Director of the Botanisches Institüt, 6300 Geissen, West Germany. When plant identifications, under the direction of Dr. Gottsberger, have been made and the collection analyzed, more detailed data will be presented to explicate the management of tropical forests by the Kayapó Indians. The research was funded by the Wenner-Gren Foundation for Anthropological Research, the Universidade Federal do Maranhão, and the Conselho Nacional de Pesquisas (CNPq). Further details on the ethnobiological study of the Kayapó Indians can be obtained by writing to Dr. Posey, Laboratório de Etnobiologia, a/c Depto. de Biologia, Universidade Federal do Maranhão, 65.000 São Luís, Maranhão, Brazil.

Literature Cited

Alcorn, J. B. 1981. Huastic noncrop resource management: Implications for prehistoric rain forest management. Human Ecology 9(4): 395–417.

Barbira-Scazzocchio, F., editor. 1980. Land, people, and planning in contemporary Amazônia. Occasional Publication 3. Cambridge University Centre for Latin American Studies, Cambridge, U.K.

Cavalcante, P. 1972. Frutas comestíves da Amazônia. Vol. 1. Publicações Avulsas do Museu Goeldi, Belém, Brazil.

———. 1974. Frutas comestíves da Amazônia. Vol. 2. Publicações Avulsas do Museu Goeldi, Belém, Brazil.

———. 1979. Frutas comestíves da Amazônia. Vol. 3. Publicações Avulsas do Museu Goeldi, Belém, Brazil.

Denevan, W. 1976. The native population of the Americas in 1492. University of Wisconsin Press, Madison.

Dobyns, H. F. 1966. Estimating aboriginal American population. Current Anthropology 7: 395–416.

Hecht, S. B. 1983. Cattle ranching in the Eastern Amazon: Environmental and social implications. *In:* E. Moran, editor. The dilemma of Amazonian development. Westview Press, Boulder, Colorado.

Kerr, W., D. A. Posey & W. Wolter Filho. 1978. Cupá, ou cipó babão, alimento do alguns índios amazônicos. Acta Amazônica 8(4): 702–705.

Moran, E. F., editor. 1983. The dilemma of Amazonian development. Westview Press, Boulder, Colorado.

Posey, D. A. 1982. The keepers of the forest. New York Botanical Garden Magazine 6(1): 18–24.

———. 1983a. Indigenous ecological knowledge and development of the Amazon. *In:* E. Moran, editor. The dilemma of Amazonian development. Westview Press, Boulder, Colorado.

———. 1983b. Keeping of stingless bees by the Kayapó Indians of Brazil. Journal of Ethnobiology 3(1): 63–73.

———, **J. Frechione, J. Eddins & L. Francelino da Silva.** In press. Ethnoecology as applied anthropology in Amazonian development. Human Organization.

Smith, N. 1977. Human exploitation of terra firma fauna in Amazônia. Ciência e Cultura 30(1): 17–23.

———. 1980. Anthrosols and human carrying capacity in Amazônia. Annals of the Association of American Geographers 70(4): 553–566.

Turner, T. 1966. Social structure and political organization among the northern Cayapó. Ph.D. dissertation, Department of Social Relations, Harvard University, Cambridge, Massachusetts.

The Use of Edible Fungi by Amazonian Indians

Ghillean T. Prance

The fungi are a forest resource that is often overlooked in ethnobotanical studies. Fungi are not often eaten by Amazonian Indians, but as we shall see they can be an important resource for some tribes. Fidalgo (1965, 1968) reviewed the knowledge of the ethnomycology of the Brazilian Indian tribes and concluded that they are not a mycophilous people. He cited very few examples of the consumption of fungi such as the mention made in 1823 by Spix and Martius (1938) of the use of edible fungi by the Maués Indians. Fidalgo also identified a Roquete-Pinto (1917) specimen of fungus (*Gloeoporus conchoides* Mont.) eaten by the Nhambiquara Indians. Fidalgo expressed surprise, however, at the lack of use of fungi in the region. Although this is certainly true for the majority of Indian tribes, the Yanomamo are an exception and eat many fungi. This paper is mainly a summary of my work with the Yanomamo.

It is also interesting that the Amazonian Indians do not appear generally to use hallucinogenic fungi although some species occur in the region. I have heard of only one case of use of psychoactive fungi in lowland tribes, that among the Warani (W. Davis, pers. comm.). Lowland Indians have developed their hallucinogens from a wide variety of other plants such as species of *Virola, Banisteriopsis,* and *Anadenanthera* (Prance, 1972).

Fidalgo and Hirata (1979) presented new ethnomycological data from three tribes gathered since the earlier reports of the first author. The Caiabi Indians use an unidentified Polyporaceae which they call "uepo-mutab." "Uepo" is their term for any bracket fungus, edible or not. The Txicão Indians use the term "apco" for edible fungi and "apcon" for non-edible ones. They eat *Lentinus crinitus* (L. ex Fr.) Fr. ("apco-taguo") and *Auricularia fuscosuccinea* (Mont.) Farlow ("apco-pilao"). The Txucarramãe Indians eat fungi only in the case of extreme hunger. They call all edible fungi "pinhamak" which they dry in the sun and then roast before eating. They eat *Trametes cubensis* (Mont.) Sacc., *Pycnoporus sanguineus* (L. ex Fr.) Murr., *Trichaptum trichomallum* (Berk. & Mont.) Murr., and *Auricularia fuscosuccinea* (Mont.) Farlow.

A considerable number of edible fungi are used in Zaire, Africa (Thoen et al., 1973; Parent & Thoen, 1978). They found that the protein content was relatively

Table I
The edible fungi of the Sanama Yanomamo Indians at Auaris

Sanama name	Etymology	How eaten[a]	Scientific name	Collection numbers (Prance et al.)	Habitat[b]
ADABAMO			Favolus brunneolus Berk. & Curt.	21318	
ATAPA-AMO		B	Favolus tesselatus Mont.	20082, 21329	MP
COINI-AMO	coini (=hairy) + amo	B	Lentinus crinitus (L. ex Fr.) Fr.	20024, 21315, 21355	MP
COROBAMO (=CODOBAMO, COROBO-AMO, COTO-AMO)	corob (=chest) + amo	B	Polyporus tricholoma Mont.	21313	MP
HAMIMAMO	hami (=peppery tasting, like Capsicum) + amo	B	Lentinus sp.	voucher lost	
			Pleurotus sp.	20085, 21326	MP
HAMIMAMO-AMWAI	hami (as above) + amo + wai (=small)	B	Lactocollybia aequatorialis Sing. (Fig. 1)	21414	MP
HASSAMO	hassa (=deer) + amo	B	Polyporus sp.	21332, 21447	
			Favolus striatulus Ellis & Ev.	21501	
HIWALAMO	hiwala (=porcupine) + amo	B	Pleurotus sp.	21330, 21510	
I-NISHI-AMO (=I-NISHI-MI-AMO)	nishi (=small) + amo	B	Pholiota bicolor (Speg.) Sing. (Fig. 2)	21322	SF
NAI-NAI-AMO		B, C	Lentinus glabratus Mont. in Sagra	20084, 21328	MP
PIDA-PIDA-LHAMO[c]			Gymnopilus hispidellus Murr.	21550	
PLO-PLO-LEMO-AMO (=PLO-PLO-KE-AMO, PO-PO-LE-AMO)	plo-plo (=the sound made by a toad) + amo	B	Pleurotus concavus (Berk.) Sing.	20088, 21331	MP
SAMA-SAMA-IAMO	sama-sama (=sting-ray) + amo	C	Polyporus aquosus Henn.	21316	F

Table I (continued)

Sanama name	Etymology	How eaten[a]	Scientific name	Collection numbers (Prance et al.)	Habitat[b]
SHI-KIMO-AMO	shi-kima (=small parrot) + amo	B	*Coriolus zonatus* (Nees) Quélet	21398, 21416	MP
SHI-KIMO-AMO-QUE	shi-kima (as above) + amoque	B	*Hydnopolyporus palmatus* (Hook. in Kunth) O. Fid.	20083, 21397, 21576	MP
SHIO-KONI-AMO (=SHI-KEMA-AMO-QUE)	shio (=anus) + coini (=hairy) + amo	B	*Panus rudis* Fr.	20016, 21327, 21333	MP
		B	*Lentinus crinitus* (L. ex Fr.) Fr.	20015, 21334	MP
		B	*Lentinus velutinus* Fr. (Fig. 3)	20016, 21392	MP
WAIKASSAMO	waika (=the people) + amo	B, R	*Favolus brasiliensis* (Fr.) Fr. (Fig. 4)	20014, 21314, 21317	MP, F

[a] B = boiled, C = raw, R = roasted.
[b] F = forest, MP = manihot plantation or old field, SF = secondary forest.
[c] Some Indians said this species is eaten, others said no.

Table II

The edible fungi collected at the Yanomamo Indian village in the Serra das Surucucus

Indian name	Scientific name	Prance collection number
SHIKIMAMOK	*Favolus brasiliensis* (Fr.) Fr.	10526, 13602
MAFCOMKUK	*Polyporus stipitarius* Berk. & Curt.	10515
ADAMASIK	*Favolus tessellatus* Mont.	13615
HODOHKUK	*Neoclitocybe bissiseda* (Bres.) Sing.	10516

high ($\bar{x} = 22.7\%$) and concluded that mushrooms are an important food supplement for local populations.

The consumption of fungi by the Yanomamo has been reported in the popular literature in several places. For example, the missionary Father Cocco (1972) mentions at least three under the names "atama-asi," "korori-una," and "moxiririwe," all names quite different from those which I recorded. Donner (1982), in her book about the Yanomamo, mentioned the importance of the edible, white mushroom, called "sikomasik," growing on decaying tree trunks. Further information about the Yanomamo Indians can be found in Chagnon (1968) and Smole (1976).

Materials and methods

This study is based primarily on a visit to three different sites of Brazilian Yanomamo, where information was gathered from the users of the fungi.

The sites visited were 1) Auaris, Roraima Territory, Brazil, 65°25'W; 4°6'N; 2) Serra das Surucucus, Roraima Territory, Brazil, 63°41'W; 2°58'N; and 3) Tototobi, Amazonas, Brazil, 63°37'W; 1°47'N. Auaris and Tototobi are on the northern and southern extremes, respectively, of Brazilian Yanomamo territory and the Serra das Surucucus lies midway between the other two study sites. The people of these three villages speak different dialects of Yanomamo. Those of Auaris speak the Sanama dialect and those of Tototobi the Yanomam dialect.

We were impressed with the consistency of common names of fungi given to us by different Indians, and, therefore, believe that nomenclature for fungi is common knowledge in the tribes and that our reports are reliable. The Indians distinguish by common name botanical species which look quite similar to the non-mycologist. On one of the earlier trips we used men informants. After considerable time with the Yanomamo, we learned that it is the women, not the men, who have the greatest mycological knowledge. Therefore, we then exclusively used women as informants and a female anthropologist as interpreter.

Herbarium collections of the fungi discussed were made at all three sites, and specimens are deposited in the herbaria of INPA, NY and SP.

Results

From Tables I–III it can be seen that 21 species of edible fungi were collected at Auaris, 4 at Surucucus, and 19 at Tototobi (the latter require further identification). The sample from Surucucus is far from complete and represents the four species collected when I first realized the importance of fungi to the Yanomamo.

The samples from Auaris and Tototobi are more complete because the study of fungi was the primary goal of later field trips to those areas. In all three study sites, members of both the Polyporaceae and Agaricaceae are eaten, but species of the former family are by far the most frequently used. Only *Favolus brasiliensis* was collected at all three sites.

There is no doubt that edible fungi are an important dietary supplement for the Yanomamo, because a considerable variety of species are used regularly. In addition to plantain and wild game which are their principal food items, the Indians use many supplementary dietary items. It is the smaller quantities of a wide variety of items that gives them an exceptionally good diet. For example, in addition to fungi, they eat many fruits, insects, large beetle larvae, frogs, and the heart of a species of *Heliconia* similar to heart of palm. At Tototobi the Indians have two words for eating, one for meat and one for other dietary items. The word for meat-eating is also applied to fungus-eating, perhaps implying that they regard this source of protein as a substitute for meat.

At Auaris the majority of the fungi are boiled in water before eating, whereas at Tototobi they are more frequently roasted in a banana leaf. Most of the fungi are rather tasteless to the western palate, but one species of *Lentinus* and one of *Pleurotus* eaten at Auaris and one of *Collybia* at Tototobi are hot, like *Capsicum* peppers. The large, soft fungus *Polyporus aquosus,* bread-like in consistency and quite tasteless, is eaten raw. Some of the fungi such as *Lentinus crinitus* are tough and leathery, and it is difficult to understand why they are valued as food items. This species seems to have rather wide use since it was one of those reported by Fidalgo and Hirata (1979) for the Txicão Indians of southern Amazonia.

Both groups studied extensively (those at Auaris and Tototobi) have a system of nomenclature for their edible fungi. At Auaris the suffix "-amo" (with a few exceptions) is applied. This suffix is also applied to a few other items of food, for example, the edible apical shoot of a *Heliconia* (*P20027*) which resembles palm-heart in taste and appearance.

At Tototobi "-amok" is suffixed to the names of edible fungi. However, at least one name ends in "-kuk," as do two out of four at Surucucus. Many of the names are derived from the resemblance of the fungi to familiar items such as deer, porcupine, stingray, etc. Different species which look alike are sometimes differentiated in terms of human relationships such as that of father and son. For example, the species known as "ubixilima-amok" is distinguished from that known as "ubixilima-amok-ihiluba," the last word being the Yanomam for son.

There is no such elaborate terminology for non-edible fungi, and the majority of these are simply called "wonshela-de" at Auaris which means "no good." They are nearly all called "bolibolikuk" at Tototobi. "Boli" is the word for moon and "bolibolikuk" is applied to all non-edible Polyporaceae growing on logs and also to some Agaricaceae. Both the Auaris and Tototobi Indians applied their terms for non-edible fungi to all *Auricularia* species for which we asked names. The *Auricularia* species which are eaten by the Txicão (Fidalgo & Hirata, 1979) are eaten also in many other parts of the world such as China. The Yanomamo do not consider *Auricularia* a source of food.

The use of abandoned swidden

One of the most interesting aspects of the Yanomamo consumption of fungi is the ecological conditions under which the fungi grow. Most of the fungi used are species of Polyporaceae which grow primarily on dead wood (Figs. 5 & 6). The process of swidden agriculture creates an ideal situation for the growth of fungi

Table III

The edible fungi of the Yanomamo Indians at Tototobi

Yanomam name	Etymology	How eaten[a]	Scientific name	Prance collection numbers & habitat[b]
ALA-AMOK		B, R	*Gymnopilus earlei* Murr.	23606 (F), 23607 (F)
ALA-AMOK-A-YAY	a-yay (=true)	B	*Favolus brasiliensis* (Fr.) Fr.	23605 (F)
ALA-AMOK-Y-AE		R		23682 (P)
ALA-AMOK-DALE-OWI	dale (=spoiled) + owi (=kind)	R		23615 (F)
BOKALA-AMOK-SIK		R		23610 (F), 23653 (P), 23688
BOKALA-AMOK-SIK-HWÖ-É	hwoe (=father)	R		23652
BROKEM AMOK		R	*Leucocoprinus cheimonoceps* (Berk. & Curt.) Sing.	23663 (F)
DUHLE-AMOK		R		23650 (P)
HAYA KASI		B, R	*Lentinus* sp.	23608 (F), 23645 (P), 23646 (P)
HLAMI-LIMA-AMOK	hlami (=peppery tasting, like *Capsicum*)	R	*Collybia subpruinosa* (Murr.) Dennis	23662 (F), 23676 (P)
MAHE-KOMO-KUK		R		23611 (F), 23625 (P), 23655 (P), 23656 (P)
MOKA EMIK		R		23648 (P)
NAYNAMO-AMOK		R	*Collybia pseudocalopus* (Henn.) Sing.	23612 (F)
NAYNAMO-AMOK		R	Polyporaceae	23647 (P), 23680 (P), 23681 (P)

Table III (continued)

Yanomam name	Etymology	How eaten[a]	Scientific name	Prance collection numbers & habitat[b]
UBIXILIMA-AMOK		R		23649 (P)
UBIXILIMA-AMOK-HWÖ-É	hwoe (=father)	R		23614 (P), 23651 (P)
UBIXILIMA-AMOK-IHILUBA	ihiluba (=son)	R		23613 (F)
UBIXILIMA-AULIMA-AMOK		R		23664 (F)
UXI-LIMA-AMOK		R		23674 (P)

[a] B = boiled, R = roasted.
[b] F = field, P = plantation.

FIGS. 1–2. 1. Hamimamo-amwai (*Lactocollybia aequatorialis*) used by the Sanama at Auaris, growing on a dead log in a cassava plantation. 2. I-nishi-mi-amo (*Pholiota bicolor*) eaten by the Sanama at Auaris, growing on a tree stump in the forest.

FIGS. 3–4. 3. Shio-koni-amo (*Lentinus velutinus*) one of the edible fungi of the Sanama. 4. Wai-kassamo (*Favolus brasiliensis*), the most popular edible fungus of the Yanomamo which was collected at all three villages.

on the many logs which are left lying in the fields. In typical banana or *Manihot* plantations, the crop grows in gaps between the network of felled trees. When the rains begin and the crops flourish, the fungi also begin to grow on the logs. The Indians are therefore inadvertently cultivating a second crop, the fungi. Many of the fungi, such as *Favolus brasiliensis* (Fig. 4) and *Polyporus tricholoma,* listed in Tables I–III are characteristic of rotting logs. The fungi flourish under the combination of humid climate and an abundance of fallen logs, pieces of half-burnt, rotting wood, and standing tree trunks.

The standing trunks and fallen logs remain long after the plots have been 'abandoned' for their primary crops. Usually the first kind of place to which we were led by the Indians at both Tototobi and Auaris to collect edible fungi was abandoned fields. It is becoming increasingly apparent that such secondary fields play an important role in the subsistence of many groups, such as the Kayapó (Posey, 1982, 1983) and the Bora Indians of Peru (Padoch, pers. comm.), long after the fields are no longer used for the initial crop. Chagnon and Hanes (1979) pointed out the importance of fallow fields as a source of food for the Yanomamo, and Harris (1971) also observed this in the Orinoco region of Venezuela. Provision from so-called fallow areas is discussed also in some detail by Kunstadter (1978) for the Lua' swiddeners in northwestern Thailand. The Yanomamo Indians at all three studied villages make frequent visits to fallow fields to gather fungi. They obtain the majority of their fungi from active and abandoned fields rather than from the forest. Because the women visit the fields daily to weed and harvest the crops, they have the knowledge of edible fungi. Gathering fungi is, therefore, much more compatible with their role than with that of the men who spend more time in the forest as hunters.

The use of fungal sclerotia

The large, usually subterranean mass of hyphae (sclerotium) formed by some fungi is used by native people in many parts of the world. Various groups of Indians widely scattered throughout the Amazonian region have discovered this resource. An Amazonian species *Polyporus indigenus* which produces sclerotia was recently described (Araujo Aguiar & Sousa, 1981). It is called "pão do índio" or Indian bread by the settlers of the region.

This spherical to ovoid mass can weigh well over three kilos and contains 50.5 percent carbohydrate (Araujo Aguiar & Sousa, 1981). The Indians use it as an emergency food rather than as a regular dietary item. The caboclos, settlers of the region, also eat sclerotia. Strips are cut off the sclerotium and either boiled, fried or put into a stew.

The sclerotium of *Polyporus saporema* Möller is also used in Brazil and is also called Indian bread (Maravalhas, 1965). Sclerotia were called Indian bread before their fungal origin was realized. Early accounts tell how the Indians baked a special type of bread and buried it in the soil to await their needs on a hunting trip or during a war.

The Indians are skilled at finding the sclerotia of *P. indigenus.* They occur near the surface of the soil and are felt by the Indians, as they walk through the forest, as hard, rock-like areas underfoot. When they feel a sclerotium they will stop and dig it up. The sclerotia are a convenient form of food because, as long as they are kept dry, they remain usable for many months. They decompose and produce fruiting bodies only in warm humid atmosphere.

Oso (1977) recorded that the sclerotia of *Pleurotus tuber-regium* (Fr.) Sing. are

FIGS. 5–6.   5. Yanomamo clearing showing the quantity of dead logs which remain in a field as suitable fungal substrate. 6. Edible fungi of Sanama at Auaris growing on logs scattered among the cassava plants.

eaten in Nigeria and are of considerable economic importance both as a food and in traditional medicine. Their use is also well known among North American Indians who call them "tuckahoes." Weber (1929) gave a detailed account of the use of sclerotia of *Poria cocos* (Schw.) Wolf, and many earlier writers recorded their uses. For example, Rafinesque (1830) wrote that tuckahoes were "the most delicate of all foods, inodorous and of fine taste." In North America they were generally prepared by roasting.

Acknowledgments

I thank Drs. Rolf Singer and Oswaldo Fidalgo for the identification of the fungi and the Unevangelized Field Mission, Northwest Amazon and the New Tribes Mission for permission to use their mission stations at Auaris and Tototobi, respectively. I thank the Instituto Nacional de Pesquisas da Amazônia and the Fundação Nacional do Indios for support and permission to work in Brazil. My special thanks go to anthropologist Alcida Rita Ramos and Pastor Donald Borgman for acting as interpreters at Tototobi and Auaris. Field work was supported by National Science Foundation Grants GB-32575X3 and INT 77-17704.

Literature Cited

Araujo Aguiar, I. de J. & M. A. de Sousa. 1981. *Polyporus indigenus* I. Araujo & M. A. Sousa, nova espécie da Amazônia. Acta Amazônica 11: 449–455.

Chagnon, N. A. 1968. Yanomamö. The fierce people. Holt, Rinehart and Winston, New York.

———— **& R. B. Hanes.** 1979. Protein deficiency and tribal warfare in Amazonia: New data. Science 203: 910–913.

Cocco, P. L. 1972. Iyëwei-teri, quince años entre los yanomamos. Escuela Técnica Don Bosco Boleíta, Caracas, Venezuela.

Donner, F. 1982. Shabono. Delacourte Press, New York.

Fidalgo, O. 1965. Conhecimento micológico dos índios brasileiras. Rickia 2: 1–10.

————. 1968. Conhecimento micológico dos índios brasileiros. Rev. Antrop. 15–16: 27–34.

———— **& J. M. Hirata.** 1979. Etnomicologia Caiabi, Txicão e Txucarramãe. Rickia 8: 1–5.

Harris, D. R. 1971. The ecology of swidden cultivation in the upper Orinoco rain forest, Venezuela. Geographical Rev. 61: 475–495.

Kunstadter, P. 1978. Ecological modification and adaptation: An ethnobotanical view of Lua' swiddeners in Northwestern Thailand. Pages 169–200. *In:* R. I. Ford, editor. The nature and status of ethnobotany. Mus. of Anthrop., Univ. of Michigan, Anthropology papers No. 67. Ann Arbor, Michigan.

Maravalhas, N. 1965. O pão de índio. Ciência e Cultura 17: 18–20.

Oso, B. A. 1977. *Pleurotus tuber-regium* from Nigeria. Mycologia 69: 271–279.

Parent, G. & D. Thoen. 1978. Food value of edible mushrooms from Upper-Shabu region. Econ. Bot. 31: 436–445.

Posey, D. A. 1982. The keepers of the forest. Garden 6(1): 18–24.

————. 1983. Indigenous knowledge and development: an ideological bridge to the future. Ciência e Cultura 35: 877–894.

Prance, G. T. 1972. An ethnobotanical comparison of four tribes of Amazonian Indians. Acta Amazônica 2(2): 7–27.

Rafinesque, C. S. 1830. Medical flora of North America 2: 270.

Roquete-Pinto, E. 1917. Rondônia. Arch. Mus. Nac. Rio de Janeiro 20: 1–252.

Smole, W. J. 1976. The Yanoama Indians: A cultural geography. Univ. of Texas Press, Austin.

Spix, J. B. von & C. F. P. von Martius. 1938. Viagem pelo Brasil. Vol. 1. Impressa Nacional, Rio de Janeiro, Brazil (Portuguese translation of 1823 publication).

Thoen, D., G. Parent & T. Lukengu. 1973. L'usage des champignons dans le Haut-Shaba. Problèmes Sociaux Zaïrois 100–101: 69–85.

Weber, G. F. 1929. The occurrence of tuckahoes and *Poria cocos* in Florida. Mycologia 21: 113–130.

Ver-o-Peso: The Ethnobotany of an Amazonian Market

Maria Elisabeth van den Berg

Ver-o-Peso is the traditional name of a large, open-air market union in Belém, State of Pará, in the Amazonian Region of Brazil. It is situated at the boundary between the old and the commercial parts of the city and near the ancient Fort of Castelo on the waterfront of the confluence of Guamá, Moju, and Acará Rivers that form Guajará Bay. The adjacent dock of Ver-o-Peso allows a variety of boats to bring in the regional products that provide the people of Belém and vicinity. In addition to its economic importance, the market has much folkloric, touristic, and ethnobotanical interest. There have been no ethnobotanical studies of this market although it is old and well known.

I began to study the market in 1965 when I initiated the collection of data and increased the intensity of my study in 1970 when I began a detailed study of the medicinal plants.

Voucher collections from the Ver-o-Peso are deposited in the herbarium of the Museu Goeldi (MG) and were identified there by myself or by recognized specialists. I consulted current taxonomic literature and used specimens which were reliably identified by specialists for comparison and avoided the consultation of general dictionaries and checklists. Specimens of dubious identity were sent to specialists (for example, to Kubitzki for *Caraipa* and Prance for Chrysobalanaceae and Lecythidaceae). Most of the medicinal plants listed here for the Ver-o-Peso are cited with more detail in Berg (1982).

I found that the species sold in this market are native to many places throughout the Amazon Basin and are also found in northeastern and southern Brazil. In addition, many introduced exotic species are sold. Some species that are common also in western Amazonia are cited in Silva et al. (1977) under other local names. Species originating from the Federal Territory of Amapá are cited in Rabelo and Berg (1981). For cultivated and exotic plants, the Latin binomials cited by Zeven and Zhukovsky (1975) are used here.

In the market from the dock towards the Praça do Pescador (Fisherman's Square), five distinct sections of the market can be found: 1) the handicraft market, 2) the medicinal and magic plant market, 3) the vegetable and root crop market, 4) the fruit market, and 5) the horticultural and ornamental plant market.

Advances in Economic Botany 1: 140–149, 1984
© 1984 The New York Botanical Garden

In this paper I have listed the most common plants encountered in each section of the market. The names are listed alphabetically by the common names as they are used and spelled in Belém and are followed by their scientific names, family names, and brief annotations on uses where appropriate.

The handicraft market

Balata—*Manilkara bidentata* (A. DC.) A. Chev. [=*Ecclinusa balata* (Ducke) Baehni or *Chrysophyllum balata* Ducke] (SAPOTACEAE). Rubber-like latex is used to make toys.

Caraipé—*Licania sclerophylla* (Mart. ex Hook. f.) Fritsch, *L. octandra* (Hoffm. ex R. & S.) Kuntze, *L. rigida* Benth. and *Licania* spp. (CHRYSOBALANA-CEAE). The ash from the bark is used in potter's clay to make ceramic objects harder and more heat resistant.

Cipó titica—*Heteropsis spruceana* Schott (ARACEAE). Cane from aerial roots is used for baskets.

Coqueiro—*Cocos nucifera* L. (PALMAE). Fibers are used for handicrafts.

Cuia—*Crescentia cujete* L. (BIGNONIACEAE). Calabashes, the most-used utensil (Fig. 1).

Guarumá—*Ischnosiphon ovatus* Koern. and *I. gracilis* (Rudge) Koern. (MAR-ANTACEAE). Stem is split and the cane is used for baskets, sieves, and fire-fans (Fig. 2).

Inajá—*Maximiliana regia* Mart. (PALMAE). The leaves provide excellent cane material for several kinds of sieves.

Jupati—*Raphia taedigera* Mart. (PALMAE). The leaf base is employed as spits and shrimp or "siri" fish equipment.

Miriti—*Mauritia flexuosa* L. f. (PALMAE). The petiole is used chiefly in the manufacture of toys.

Ubuçu—*Manicaria saccifera* Gaertn. (PALMAE). The fibrous material around the leaf bases is used for handicrafts.

The medicinal and magic plant market

In this area (Fig. 3) there are preparations of both plant and, to a lesser extent, animal origin which are utilized in folk medicine or in the magic rites of Afro-Brazilian religious cults. The stalls sell whole herbs, bark, roots, leaves, seeds, resins, and inflorescences that are used as teas, infusions, ointments, incense, bath salts, and smoking material and in many other ways. The people of Belém use many scented plants as perfumes in their rituals and to attract members of the opposite sex. Such plants are referred to in the following list as attractants. The principal species available are:

Alecrim—*Rosmarinus officinalis* L. (LABIATAE). Leaves are aromatic, and medicinal when inhaled.

Amapá—*Parahancornia amapa* (Huber) Ducke (APOCYNACEAE). Latex is used as a medicinal tonic.

Amapá-dôce—*Brosimum potabile* Ducke (MORACEAE). Latex is used as a medicinal tonic (Fig. 3).

Anador—*Coleus barbatus* Benth. (LABIATAE). Leaves are used to treat digestive disturbances.

Ananaí—*Ananas ananassoides* (Baker) L. B. Smith (BROMELIACEAE). Fruit is used as an abortifacient.

Figs. 1–3. 1. Calabash utensils in the handicraft market. 2. Flour sieves, made of *Ischnosiphon*, and calabash utensils in the handicraft market. 3. A stall in the medicinal and magic plant market. The white bottle contains latex of "Amapa doce" (*Brosimum potabile*). Other bottles contain sap of jatoba, oil of andiroba, and oil of copaifera.

Andiroba—*Carapa guianensis* Aubl. (MELIACEAE). The oil pressed from the fruit is used as an antiphlogistic and an antiarthritic.

Apií—*Dorstenia asaroides* Gard. (MORACEAE). Root tubercles are used to treat coughs.

Arruda—*Ruta graveolens* L. (RUTACEAE). Leaves are used in Afro-Brazilian rituals and as an abortifacient.

Barbatimão—*Stryphnodendron barbadetiman* (Vell.) Martins (LEGUMINOSAE-MIMOSOIDEAE). Bark of this medicinal plant is imported from the "cerrados" of Mato Grosso. I have used the nomenclature of Martins (1972) rather than the commonly applied name *S. adstringens* (Mart.) Cov.

Baunilha—*Vanilla* spp. (ORCHIDACEAE). An aromatic.

Breu branco—*Protium heptaphyllum* (Aubl.) March. (BURSERACEAE). The resin is used as an aromatic in ritual incense.

Cabacinha—*Luffa operculata* (L.) Cogn. (CUCURBITACEAE). Fruit when mixed with "pião branco" is used to treat sinusitis.

Canarana—*Costus spicatus* Rosc. (ZINGIBERACEAE). Leaves are used to treat kidney problems.

Canela—*Cinnamomum zeylanicum* Breyn. [=*Laurus cinnamomum* L.] (LAURACEAE). Leaves and bark are aromatic and used as a carminative; they are also used as an attractant in rituals.

Catuaba—*Anemopaegma arvense* (Vell.) Stel. (BIGNONIACEAE). The entire plant is used as an aphrodisiac. Also *Erythroxylum* cf. *vacciniifolium* Mart. (ERYTHROXYLACEAE), *Trichilia* sp. (MELIACEAE), *Pouteria* sp. (SAPOTACEAE) and a species of BURSERACEAE. The barks of these latter species are all used as aphrodisiacs and also in an infusion with "marapuama" and "gengibre."

Cipo d'alho—*Adenocalymma alliaceum* Miers (BIGNONIACEAE). Leaves are used in ritual baths and incense.

Copaíba—*Copaifera reticulata* L. and *Copaifera* spp. (LEGUMINOSAE-CAESALPINIOIDEAE). The oil from the trunk is used as an anti-inflammatory and the bark is used to cicatrize gastric ulcers.

Corrente—*Pfaffia glomerata* (Spreng.) Pederson (AMARANTHACEAE). Leaf is used to make an antihemorrhoidal medicine.

Cumaru—*Dipteryx odorata* (Aubl.) Willd. (LEGUMINOSAE-PAPILIONOIDEAE). Seeds are aromatic and used in a medicine to treat otitis.

Elixir paregórico or Óleo elétrico—*Piper callosum* Ruíz & Pavón (PIPERACEAE). Leaves are aromatic and used to treat digestive disturbances and pain.

Espada-de-Joana D'Arc—*Sansevieria trifasciata* Hort. ex Prain (AGAVACEAE). This plant is used in Afro-Brazilian rituals.

Espada-de-São Jorge—*Sansevieria guineensis* (L.) Willd. (AGAVACEAE). This plant is used in Afro-Brazilian rituals.

Fava de Jucá—*Caesalpinia ferrea* Mart. var. *cearensis* Huber (LEGUMINOSAE-CAESALPINIOIDEAE). Fruit is used as an astringent.

Gengibre—*Zingiber officinale* Rosc. (ZINGIBERACEAE). Root is used as an external treatment of bronchitis and rheumatism pain.

Japana branca—*Eupatorium triplinerve* Vahl (COMPOSITAE). Leaf is used to treat gastric ulcers and as an attractant in ritual baths.

Jatobá—*Hymenaea courbaril* L. (LEGUMINOSAE-CAESALPINIOIDEAE). Bark sap is used to treat bronchitis.

Lágrima de Nossa Senhora—*Coix lacryma-jobi* L. (GRAMINEAE). Seeds are used for beads, important in Afro-Brazilian religious cults.

Lança-de-São Jorge—*Sansevieria cylindrica* Boj. (AGAVACEAE). The entire plant is used in Afro-Brazilian rituals.

Macaxeira—*Manihot esculenta* Crantz (EUPHORBIACEAE). The tuber rootlets of this edible plant are used to compound a type of ritual incense.

Maçaranduba—*Manilkara huberi* (Ducke) A. Chev. (SAPOTACEAE). The latex is used as a tonic.

Mangericão—*Ocimum micranthum* Willd. (LABIATAE). Leaves are used to treat gastric disturbances.

Mangerona—*Majorana hortensis* Moench. (LABIATAE). Leaves are used to treat cough.

Marapuama—*Ptycopetalum olacoides* Benth. (OLACACEAE). Roots are used as an aphrodisiac.

Marupazinho or Marupá-í—*Eleutherine plicata* Urban (IRIDACEAE). Bulbs are used as an antidysenteric, a most effective treatment for amoeba.

Mastruço or Mastruz—*Chenopodium ambrosioides* L. (CHENOPODIACEAE). Leaves are used as a tonic in a mixture with milk or as an anthelmintic.

Mucura-caá or Guiné—*Petiveria alliacea* L. (PHYTOLACCACEAE). Young branches are used to treat toothache and the leaves are employed in ritual baths and amulets.

Mururé-da-terra-firme—*Brosimum acutifolium* Huber (MORACEAE). Bark and especially roots are used as an antirheumatic.

Paricá—*Piptadenia peregrina* (L.) Benth. (LEGUMINOSAE-MIMOSOIDEAE). Bark used as a cicatrizant and in ritual incense.

Pariri or crajirú—*Arrabidaea chica* (H.B.K.) Verlot (BIGNONIACEAE). Leaves are used as an anti-inflammatory or a tonic.

Pataquiera—*Conobea scoparioides* Benth. (SCROPHULARIACEAE). Leaves are aromatic.

Patichuli—*Vetiveria zizanioides* (L.) Stapf (GRAMINEAE). The popular nomenclature of this species in Pará is confusing because here *Vetiveria* is called "patichuli" causing confusion with true "patichuli" which is *Pogostemon* spp. (which is named "oriza" or "uriza" in Pará). *Vetiveria* is named "vetiver" in other localities. The aromatic roots are used as a medicine and as an attractant.

Pau rosa—*Aniba rosaeodora* Ducke and *A. duckei* Kosterm. (LAURACEAE). The bark and wood are aromatic, and used as attractants.

Pião branco—*Jatropha curcas* L. (EUPHORBIACEAE). The plant is used in Afro-Brazilian cult rites, when mixed with "cabacinha," and the latex is used as a medicine for sinusitis. However, practitioners of folk medicine advise caution because the latex is very caustic and needs careful preparation.

Pião roxo—*Jatropha gossypifolia* L. (EUPHORBIACEAE). Plant is used in Afro-Brazilian rituals.

Pirarucu or Folha-da-fortuna—*Bryophyllum calycinum* Salisb. (CRASSULACEAE). Leaves are used as an antiphlogistic and an emollient. They were used with great success during an outbreak of conjunctivitis in Belém.

Priprioca—*Cyperus odoratus* L. (CYPERACEAE). Root is aromatic and used as an attractant.

Pucá—*Cissus sicyoides* L. (VITACEAE). Leaves used as a hypotensor.

Sacaca—*Croton cajucara* Benth. (EUPHORBIACEAE). Bark and leaves are aromatic and used to treat digestive disturbances.

Sucuriju—*Mikania lindleyana* DC. (COMPOSITAE). Leaves used as a cicatrizant for gastric ulcers.

Sucuúba—*Himatanthus sucuuba* (Spruce ex M. Arg.) Woodson (APOCYNACEAE). Latex is used as an anti-tumor agent, and the bark is used to treat gastric ulcers.

Tajás—*Caladium* spp. and *Philodendron* spp. (ARACEAE). Plant is used in Afro-Brazilian rituals.

Tamanquaré—*Caraipa densifolia* Mart. (GUTTIFERAE). It is the most frequently used species of *Caraipa*—the leaves as an aphrodisiac and the bark and latex to treat herpes and tetter.

Uriza or Oriza—*Pogostemon heyneanus* Benth. (LABIATAE). Leaves are aromatic and used when washing clothes to make them sweet-scented.

The vegetable and root crop market

This area of the market contains the retailers of vegetables, cassava, flours (farinha), legumes, root crops, and other edible products (Figs. 4 & 5). Here, attractive clay pots containing "melado" (boiled sugar cane juice) are sold. The pots are closed with "guarumá" (*Ischnosiphon*) leaves and packed in a cane basket made from the stems of guarumá. Some vegetables and other products available

FIGS. 4–5. 4. Lettuce in the vegetable market. 5. Vegetables laid out on the ground in the vegetable market. The piles of herbs on *Ischnosiphon* leaves are "cheiro verde" as described in the text. The round, spinous fruit is "maxixe," and in the foreground is "quiabo" (okra).

in this section reflect the recent wave of Japanese immigration to Northern Brazil. They practice intensive horticulture and market-gardening and have greatly increased the variety and quantity of vegetables available. The most important plants sold in this market are:

Alface—*Lactuca sativa* L. (COMPOSITAE). Lettuce (Fig. 4).
Alfavaca—*Ocimum gratissimum* L. (LABIATAE).
Batata doce—*Ipomoea batatas* L. (CONVOLVULACEAE). Sweet potato.
Beringela—*Solanum melongena* L. (SOLANACEAE). Eggplant.
Cará—*Dioscorea brasiliensis* Willd. (DIOSCOREACEAE). Yam.
Cebolinha—*Allium fistulosum* L. (LILIACEAE). Chives.
Cenoura—*Daucus carota* L. (UMBELLIFERAE). Carrot.
"Cheiro Verde"—This is a packet with a mixture of salsa, coentro, and cebolinha
    (Fig. 5).
Chicória—*Cichorium intybus* L. (COMPOSITAE). Chicory.
Coentro—*Coriandrum sativum* L. (UMBELLIFERAE). Coriander.
Inhame—*Dioscorea alata* L. (DIOSCOREACEAE). Yam.
Jambu—*Spilanthes oleracea* L. (COMPOSITAE). A local leaf used to flavor the
    local dishes of tucupi and tacaca.
Jerimum or Abóbora—*Cucurbita pepo* L. (CUCURBITACEAE). Pumpkin.
Maniva—*Manihot esculenta* Crantz (EUPHORBIACEAE). Cassava.
Maxixe—*Cucumis anguria* L. (CUCURBITACEAE) (Fig. 5).
Pimenta-de-cheiro—*Capsicum frutescens* L. forma (SOLANACEAE). Red pep-
    per.
Pimenta Malagueta—*Capsicum frutescens* L. forma (SOLANACEAE). Mala-
    gueta pepper.
Quiabo—*Hibiscus esculentus* L. (MALVACEAE). Okra (Fig. 5).
Salsa—*Petroselinum sativum* Hoffm. (UMBELLIFERAE). Parsley.
Tomate—*Lycopersicum esculentum* (L.) Mill. [=*Solanum lycopersicum* L.] (SO-
    LANACEAE). Tomato.

The fruit market

In this area I have observed that over 100 species are sold during the course
of the year with about 40 species available each day. The most appreciated fruits
are:

Abacate—*Persea americana* Mill. (LAURACEAE). Avocado.
Abacaxi—*Ananas comosus* (L.) Merrill (BROMELIACEAE). Pineapple.
Abricó—*Mammea americana* Jacq. (GUTTIFERAE). Mammey apple.
Açaí—*Euterpe oleracea* Mart. (PALMAE). Assai palm.
Bacaba—*Oenocarpus bacaba* Mart. (PALMAE).
Bacuri—*Platonia insignis* Mart. (GUTTIFERAE).
Banana—*Musa sapientum* L. and its varieties (MUSACEAE). Banana.
Banana chorona—*Musa paradisiaca* L. (MUSACEAE). Banana.
Biribá—*Rollinia orthopetala* (L.) A. DC. (ANNONACEAE).
Cacau—*Theobroma cacao* L. (STERCULIACEAE). Cocoa.
Caju—*Anacardium occidentale* L. (ANACARDIACEAE). Cashew.
Carambola—*Averrhoa carambola* L. (OXALIDACEAE). Carambola.
Castanha-do-Pará—*Bertholletia excelsa* Humb. & Bonpl. (LECYTHIDACEAE).
    Brazil nut.
Castanha sapucaia—*Lecythis usitata* Miers (LECYTHIDAECEAE). Sapucaia nut.
Coco—*Cocos nucifera* L. (PALMAE). Coconut.
Cupuaçu—*Theobroma grandiflorum* (Willd. ex Spreng.) Schum. (STERCULI-
    ACEAE).
Genipapo—*Genipa americana* L. (RUBIACEAE). Genipap.
Graviola—*Annona muricata* L. (ANNONACEAE). Soursop.
Inajá—*Maximiliana regia* Mart. (PALMAE).

Jaca—*Artocarpus integrifolia* L.f. (MORACEAE). Jackfruit.
Jambo—*Syzigium malaccense* (L.) Merrill & Perry (MYRTACEAE). Malay apple.
Jambo-rosa—*Eugenia jambos* L. (MYRTACEAE).
Laranja—*Citrus sinensis* (L.) Osbeck (RUTACEAE). Orange.
Lima—*Citrus aurantifolia* Swingler (RUTACEAE). Lime.
Limão—*Citrus limonia* Osbeck (RUTACEAE). Lemon.
Limão-de-Cayenna—*Averrhoa bilimbi* L. (OXALIDACEAE). Citronell.
Limãozinho—*Citrus limetta* Risso (RUTACEAE).
Mamão—*Carica papaya* L. (CARICACEAE). Papaya.
Manga—*Mangifera indica* L. (ANACARDIACEAE). Mango.
Mangaba—*Hancornia speciosa* Gomez (APOCYNACEAE).
Maracujá—*Passiflora edulis* Sims. (PASSIFLORACEAE). Passionfruit.
Melancia—*Citrullus vulgaris* Schrad. (CUCURBITACEAE). Watermelon.
Melão—*Cucurbita melopepo* L. (CUCURBITACEAE). Melon.
Mucajá—*Acrocomia sclerocarpa* Mart. (PALMAE).
Muruci—*Brysonima chrysophylla* L. (Kunth) (MALPIGHIACEAE).
Piquiá—*Caryocar villosum* (Aubl.) Pers. (CARYOCARACEAE).
Pupunha—*Bactris speciosa* (Mart.) Karst. (PALMAE). Peach palm.
Sapotilha—*Achras sapota* L. (SAPOTACEAE). Sapodilla.
Tamarindo—*Tamarindus indica* L. (LEGUMINOSAE). Tamarind.
Tangerina—*Citrus nobilis* Lour. (RUTACEAE). Tangerine.
Taperêbá—*Spondias lutea* L. (ANACARDIACEAE). Hog plum.
Tucumã—*Astrocaryum tucuma* Mart. (PALMAE). Star-nut palm.
Uchi—*Endopleura uchi* (Huber) Cuatr. (HUMIRIACEAE). Uchi.

The horticultural and ornamental plant market

The section in and around the Praça do Pescador at the end of the market farthest from the dock concentrates on the sale of ornamental house plants, mostly native species and others introduced by the Portuguese since the 18th Century. In response to the popular demand of the people of Belém many ornamental plants from other Brazilian cities are also sold. The most available species are:

Angélica—*Polianthes tuberosa* L. (AGAVACEAE).
Antúrio—*Anthurium bellum* Schott (ARACEAE).
Begonia—*Begonia palmata* DC., *B. semperflorens* Link & Otto and numerous other species of *Begonia* (BEGONIACEAE).
Beijo-de-frade—*Impatiens balsamina* L. (BALSAMINACEAE).
Bela Emília—*Plumbago capensis* Thunb. (PLUMBAGINACEAE).
Brasileirinho—*Caladium humboldtianum* Schott (ARACEAE).
Brinco-de-princesa—*Clerodendron thomsonae* Balf. (VERBENACEAE).
Bungavilia—*Bougainvillea spectabilis* Willd. (NYCTAGINACEAE).
Chifre-de-veado—*Platycerium alcicorne* Desv. (POLYPODIACEAE).
Camarão amarelo—*Beloperone guttata* T. S. Brand. (ACANTHACEAE).
Cidreirinha—*Lantana camara* L. (VERBENACEAE).
Crisântemo—*Chrysanthemum* spp. (COMPOSITAE).
Cristo-de-galo—*Celosia cristata* Moq. (AMARANTHACEAE).
Croton—*Codiaeum variegatum* L. (EUPHORBIACEAE).
Croton—*Coleus blumei* Benth. (LABIATAE).
Croton—*Cordyline terminalis* Kunth (LILIACEAE).
Dracena—*Dracaena fragrans* Ker-Gawl. (LILIACEAE).
Girassol—*Helianthus annuus* L. (COMPOSITAE).

Jasmim—*Jasminum officinale* L. (OLEACEAE).

Jasmim vermelho—*Ixora stricta* Roxb. (RUBIACEAE).

Labirinto—*Fittonia verschaffeltii* E. Coem. (ACANTHACEAE).

Laço-de-amor—*Episcia cupreata* Hanst. and *E. hirsuta* (Benth.) Hanst. (GES-NERIACEAE).

Margaridinha—*Wulffia baccata* L. (COMPOSITAE).

Maria-sem-vergonha—*Catharanthus roseus* G. Don (APOCYNACEAE). This species is used extensively as an ornamental and is a recognized antimetastatic.

Onze-horas—*Portulaca* sp. (PORTULACACEAE).

Orquídea—Numerous species of orchids are sold here, and they belong mainly to the genera *Laelia, Cattleya, Oncidium, Epidendrum,* and *Catasetum* (OR-CHIDACEAE).

Papoula—*Hibiscus rosa-sinensis* L. (MALVACEAE). Single flowers with five petals only. There are a large number of cultivars of several colors.

Papoula dobrada—*Hibiscus rosa-sinensis* L. (MALVACEAE). Flowers with multiple petals with a large number of cultivars in many colors.

Perpétua—*Gomphrena globosa* L. (AMARANTHACEAE).

Pimenta-de-salão—*Capsicum frutescens* Mill. forma (SOLANACEAE).

Renda portuguesa or Quebra vaso—*Davallia fijiensis* Hook. (DAVALLIACEAE).

Rosa—*Rosa canina* L., *R. gallica* L., and *Rosa* spp. (ROSACEAE).

Rosa comum—*Rosa centrifolia* L. (ROSACEAE).

Samambaias (Ferns)—*Nephrolepis* spp., *Polypodium* spp., *Pteridium* spp. (PTER-IDOPHYTA).

Samambaia furta-cor—*Selaginella* sp. (SELAGINELLACEAE).

Sorriso-de-Maria—*Aster* sp. (COMPOSITAE).

Tajá—Common name of several *Caladium* spp., *Xanthosoma* spp. (ARACEAE).

Terezinha—*Portulaca grandiflora* Hook. (PORTULACACEAE).

Tibuchina—*Tibouchina aspera* Aubl. and *Tibouchina* sp. (MELASTOMATA-CEAE).

Verbena—*Verbena chamaedryfolia* Juss. and *V. pinnatisecta* Shau. (VERBE-NACEAE).

Vindicá—*Alpinia nutans* Rosc. (ZINGIBERACEAE). This species, in addition to being very fragrant, is also used as a sedative.

Viuvinha—*Petrea volubilis* Jacq. (VERBENACEAE).

Observations and conclusions

During the course of this eighteen-year study (1965–1983), there has been a considerable reduction in both the number of retailers and the number of species found in the Vero-o-Peso market. Already there are about 50 percent fewer species represented than when the study was initiated. This is due to the considerable migration into the Amazonian Region by people who are from other regions where there is less plant-lore and who are unfamiliar with the local flora and also to the expansion of television programs, which reach the remote parts of the state with constant commercials for synthetic medicines, plastic utensils, and a different lifestyle. This has caused the loss of much folk culture and herbal medicine. This loss emphasizes the need to catalogue the folk culture and ethnobotany of such markets so that the indigenous knowledge is not permanently lost. It is not just the primitive tribes, but also the city markets, which are subject to acculturation. However, in recent years the loss of plant information has been somewhat slowed because of a renewed interest in the natural plant products and herbal medicines in the larger Brazilian cities such as Belém, Rio de Janeiro, and São Paulo. I have observed a significant return to the products and a definite interest in the search for and use of traditional medicines. Over the last few years there has been increased activity in Brazilian and foreign universities in anthropological, pharmacological, chemical, and ethnobotanical research related to traditional products, especially local medicines. Basic identification studies, such as the one presented in this paper, are a necessary prerequisite to the above studies. Many previous chemical and other kinds of studies of market plants have been carried out on botanical material which was identified only by common name and which was without voucher specimens and without reference to any scientific name. The present study has shown the unreliability of common names because of regional variation. Many of the names used in Belém are different from those cited by Silva et al. (1977) from the State of Amazonas. In addition, common names often include several different botanical species. Even in studies of such well-known places as the Vero-o-Peso market it is essential to collect voucher specimens to document the work thoroughly. The use of common names alone can confuse reports of herbal medicines as well as specimen identification. I hope to publish in the future a complete list of all the plants available in the fascinating Vero-o-Peso market and hope that this work will show the interesting diversity of plant material available in the largest market in Brazilian Amazonia.

Acknowledgments

I thank Dr. Ghillean T. Prance of the New York Botanical Garden for encouraging my studies and for the revision of the text and Mr. Antonio Pinheiro of Museu Goeldi, Belém for the photographs.

Literature Cited

Berg, M. Elisabeth van den. 1982. Plantas medicinais na Amazônia. Conselho Nacional de Desinvolvimento Científico e Tecnológico, Programa Trópico Úmido, Belém, Brasil.

Martins, E. M. Occhioni. 1972. Sobre a nomenclatura científica do "barbatimão" do Brasil. Leandra 2(3): 79–81.

Rabelo, B. V. & M. E. van den Berg. 1981. Nota prévia sobre o estudo dos cerrados do Amapá. Anais do XXXII Congresso Nacional de Botânica do Brasil 32: 134–140.

Silva, M. F. da, P. L. Braga Lisbôa & R. C. Lobato Lisbôa. 1977. Nomes vulgares de plantas amazonicas. Conselho Nacional de Desinvolvimento Científico e Tecnológico, Instituto Nacional de Pesquisas da Amazônia, Manaus, Brasil.

Zeven, A. C. & P. M. Zhukovsky. 1975. Dictionary of cultivated plants and their centres of diversity. Centre for Agricultural Publishing and Documentation, Wageningen, The Netherlands.

Index